POCKET GUIDE TO GARDEN BIRDS

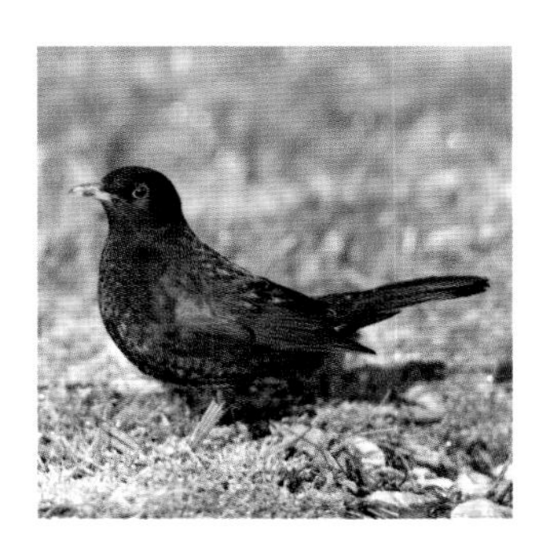

NIGEL BLAKE

BLOOMSBURY
LONDON • NEW DELHI • NEW YORK • SYDNEY

CONTENTS

Barn Swallow.

Pied Wagtail.

Blue Tit.

Mallard.

INTRODUCTION

This handy guide describes the birds that are most likely to come to gardens. The species accounts include details about how the birds live their lives, the foods that can be provided for them, and what feeders and nest boxes to use, whilst pictures of the birds ensure correct identification.

HOW GARDENS CAN BENEFIT BIRDS

However big or small your garden is it can be a haven for wildlife. Furthermore, as part of the combined wealth of habitat provided by all gardens together it can be important to the survival of many bird species, as well as other animals. It is estimated that the total area of gardens in the UK is over a million acres and increasing. This is about 3 per cent of the UK's total landmass, which might not sound like much, but is about three times more than the total area of nature reserves.

Birds as well as other creatures can benefit from how well the small patches of land that surround our homes are maintained, but the payoff for us is great too. Being able to look out of the window and see Song Thrushes and Blackbirds collecting worms for their young from the lawn, or seeing Great Tits and Blue Tits squabbling over feeders, is a joy. Seeing Spotted Flycatchers unobtrusively catching flies and using a nest box you have put up makes you feel that you are making a difference when so many species are under threat.

PROVIDING WATER

Birds need to wash and drink regularly, and in dry weather they may have to fly some distance to do so. The provision of a pool or pond in a garden soon results in birds coming to visit regularly. A small waterfall can be added to a bigger pond, since birds seem to be attracted to moving water. If you do not have the space for a pond, a birdbath can be equally useful to birds – ideally it should have a sloping bottom so that the water depth varies to suit the needs of different species.

A few rocks and twigs, or branches for small birds to perch on while they drink or bathe, make it less easy for cats to gain access to a pool, especially if the perches are well away from the edges. Tall, twiggy plants near a pool serve as places for birds to perch on as

Male Chaffinch bathing.

A Blue Tit visiting a garden bird feeder.

lookout posts before going to the pool, and as somewhere to dry off and preen after bathing. Planting some dense and perhaps thorny vegetation nearby as a safe place for birds to quickly escape to is also helpful – make sure, however, that it is vegetation which predators cannot hide in.

Most importantly, the water in small birdbaths and pools should be kept clean by being replaced regularly. In winter birdbaths and pools should be kept free from ice, as this is the time when birds depend on them most.

Swallows, House Martins, thrushes and other birds need mud for nesting, so a shallow area of mud, with some clay in it so that it will set hard, is a must for them from March to the end of May at least. Swallows and martins mix mud with bits of hay and dried grass to hold their nest structures together, so adding short, chopped dry vegetation to the mud makes it even more attractive to these birds. In a garden that is too small for a pond, the mud can be placed in a tray.

FEEDING THE BIRDS

Feeding garden birds is getting to be a popular activity, and it has turned into a profitable business that increases year on year, with high sales in winter weather. This is not just good for the companies and charities such as the RSPB that sell bird food – birds in general benefit in several ways other than just being fed.

The bird-food and feeder industry has invested in solid science and research to ensure that bird seed, dried and live insect food, and fat-cake recipes are nutritious and healthy. Food dispensers are safer for birds to use than they used to be, saving small birds from being trapped by their feet in feeders that are badly designed, or from being caught by predators while feeding. Products are also designed to suit specific species and sizes of bird, so that the food is not taken by squirrels and other rodents. Modern food containers benefit from modern materials, and are easy to dismantle for washing to keep them clean and hygienic.

Feeding sites need to be kept away from any disturbance, and should preferably be located in a place where good views of the birds can be obtained from inside the home. Instead of placing feeders close together, space them apart so that birds of various sizes and species can feed without too many conflicts. Larger garden birds such as pigeons and doves will otherwise hog the food and keep smaller birds away.

Several food mixtures designed for different types of bird are available. They are formulated to be suitable for hanging feeders, bird tables or scattering on the ground. Most include sunflower seeds, granulated peanuts and flaked maize; they can also contain wheat, barley and split peas, plus smaller seeds like millet and pinhead oatmeal, as well as dried and chopped berries. Other mixes include some of the above with fat and suet pellets, as well as dried mealworms.

Food mixes containing wheat, barley and dried peas are only suitable for birds such as pigeons and pheasants. Foods that are bulked up with lentils, dried rice and bits of biscuit should be avoided, as they are too hard for small birds to eat (biscuits are only digestible if soaked). Peanuts are best supplied in feeders with a wire mesh, so that birds cannot take whole peanuts away – these can choke young chicks. Avoid salted peanuts.

BIRD-TABLE HYGIENE

We feed birds to help them survive, and the last thing we want to do is to subject them to disease. Keeping bird tables and feeders as clean as possible is as important as choosing healthy foods. Most bird diseases are spread in their droppings, so it is essential that these are cleaned away from food areas. Points to bear in mind include:

¤ Keeping a check on the supply and demand. If food is left over reduce the amount you put out each day.

¤ Using suitable feeders and keeping them clean. Empty and remove stale or rotting food and thoroughly and regularly wash feeders, rinsing them with boiling water. Allow them to dry properly before refilling.

¤ Washing a bird table and the area around it. Use some disinfectant (4–5 per cent solution), then rinse the bird table thoroughly. If space allows regularly move the feeding site around the garden; if not dig the area around it and turn over the soil with a garden fork from time to time.

¤ Wearing gloves while you are cleaning bird-table feeders. Do this outside rather than indoors and wash your hands afterwards.

TYPES OF FEEDER

Birdfeeders are now well designed, but buying from a reputable stockist is always sensible. The RSPB has a wide range of quality products, but most garden centres are cashing in on the trend and they also have good products from several manufacturers.

Hanging feeders are available in different sizes and with a variety of perches and feeding ports. Some consist of a clear polypropylene tube with access suited to either mixed grain and seeds, or niger seeds. Others are made from plastic-coated wire, which many birds can cling to while feeding. Depending on the food that is put in them, they will attract tits, finches, Nuthatches, Treecreepers, woodpeckers and Robins.

A plastic tube feeder should have a wide-rimmed lid to stop water getting in; water combined with changes of temperature can cause condensation in the tube so that the seed or nuts inside go mouldy. Mesh feeders made with plastic-coated wire are durable, do not rust and are relatively easy to clean. They are usually designed to hold peanuts. Make sure they will not trap birds' feet and that they are strong enough to stop squirrels and Great Spotted Woodpeckers from breaking in. There are also hanging feeders made from bigger mesh for fat balls and fat cake.

Large mesh cages to fit over feeders are available. These provide protection for small birds that can fit through the mesh, enabling them to feed safely and preventing cats and Sparrowhawks from hunting them; they may also keep squirrels out.

Short-legged trays are perfect for ground-feeding species such as Dunnocks, buntings and thrushes. One with a fine-mesh bottom will drain quickly and allow the food to dry. Trays can be taken away at night to help avoid a rat problem.

Bird tables are available in a wide range of shapes and sizes. Make sure the roof is bigger than the feeding table so that the food is kept dry.

Providing for birds in a garden is a commitment, and there is some work and expense involved in feeding garden birds. When birds become used to receiving a supply of food and water in a garden, it will be the first place they will come to after roosting. If they are dependent on a feeding site they may not forage in other places, so once you start feeding birds regularly you need to keep it up. If you go on holiday arrange for a relative or neighbour to top up feeders and replace water in containers.

NEST BOXES

In recent years building regulations and local authorities' obsession with keeping everywhere tidy have resulted in a reduction in places for birds to nest – especially for small birds that nest in holes. A nest box can act as a perfect substitute for the natural holes and crevices that small birds favour. About 60 UK species use nest boxes regularly, and various box designs are available to suit their needs for warm and safe places to bring their young into the world – we gain by having a 'front-row seat' from which to watch them doing so.

Numerous designs of nest box are available. The RSPB produces several and many garden centres sell them. As well as standard boxes with a hole in the front, which are used by tits and Nuthatches, and open-fronted boxes favoured by Robins and flycatchers, there are boxes for fitting under eaves for House Martins. Big nest boxes of a type suitable for Barn Owls, Tawny Owls and Common Kestrels can be included in large gardens in suitable places.

A species that particularly benefits from help with the provision of nest boxes is the House Martin. It naturally builds cup-shaped nests with mud that sticks to the walls and soffit boards on houses. Nowadays these are likely to be made from moulded plastic, and as the nests dry out they separate from the plastic and fall down with the eggs and young.

The best time to put up nest boxes is autumn, which allows them to weather and lose the smells of glue and other materials used in their making. Birds soon investigate the boxes as shelter and roosting sites during winter, and are likely to nest in them in the following spring. To increase the chance of getting a pair of birds to move in, a next box needs to be sited appropriately, and some basic rules should be followed.

¤ Unless the site is already sheltered it is best to place a box where it will not get soaked inside by the prevailing weather. Facing it between north and east is generally best as this also avoids it being in strong sunlight during the day.

¤ All birds need a clear route in and out of their nests, with a few places to perch and check that it is safe to enter without attracting unwanted attention.

¤ Hole-fronted boxes should be positioned 2–4m up to attract tits,

Nuthatches and sparrows. For woodpeckers it is best to place a box higher up at perhaps 5–6m. Only colony-nesting birds such as Tree Sparrows use adjoining boxes. For other birds it is best to keep a good space between nest boxes (if you are using more than one box) in order to avoid conflict.

¤ Open-fronted boxes that Robins and flycatchers use are best placed on a wall or post where they are hidden by foliage; 2–4m up on an ivy-covered wall is ideal.

¤ Once birds are nesting in a box they should not be disturbed (it is in fact illegal to disturb nesting birds).

¤ Bird boxes get infested with fleas and feather mites during use so it is important to clean them out after breeding has finished. October to January is the best time to do this. All the old nesting material should be removed. It is best to wear gloves for this job, as a nest may contain unhatched eggs and dead chicks. After cleaning the box should be scalded with boiling water to kill any parasites, then allowed to dry thoroughly before the lid is replaced in preparation for new residents.

Small birds such as Robins may use a household item like an old teapot to nest in.

MALLARD

Anas platyrhynchos

This is the most common duck in the temperate northern hemisphere, a bird most of us are familiar with from a young age. The female is brown with darker mottling. The male is a splendid-looking bird with a metallic green head, white neck-ring, brown-and-grey wings, and a pale underside. Both sexes have metallic blue speculum feathers with a white surround that are obvious when they are in flight; they both also have yellow beaks. Ducklings are little brown-and-yellow bundles of fluff.

LENGTH 58cm.

HABITAT AND DISTRIBUTION Almost any patch of water across Europe.

DIET Dabbles for surface food items and upends to reach aquatic vegetation and invertebrates below the surface. Mallards do eat bread, but it is not healthy for them – grain is much better. They will come to food at garden feeding sites.

NESTING Usually nests in dense vegetation close to water. The eggs are pale blue-green with a waxy sheen; they take nearly a month to hatch.

The female Mallard is plain brown, an aid to camouflage when nesting.

The drake Mallard is brightly marked for attracting a mate.

COMMON PHEASANT

Phasianus colchicus

Pheasants were introduced to the UK, probably in the 11th century, as game birds. They breed successfully in the wild, and numbers are high due to breeding in pens by shooting estates. The male is unmistakable – a big golden-brown bird with a long tail and a metallic green or purple head and white neck-ring. The female is a brown-mottled version of the male, but is slightly smaller. The species' colours are variable.

LENGTH 75–90cm (male); 53–64cm (female).

HABITAT AND DISTRIBUTION Woodland, farmland, large rural gardens and reedbeds in much of Europe.

DIET Feeds on seeds, fruits, nuts and roots, and will take seeds in rural gardens.

NESTING Nest is a grassy cup on the ground.

This male Pheasant is displaying and calling.

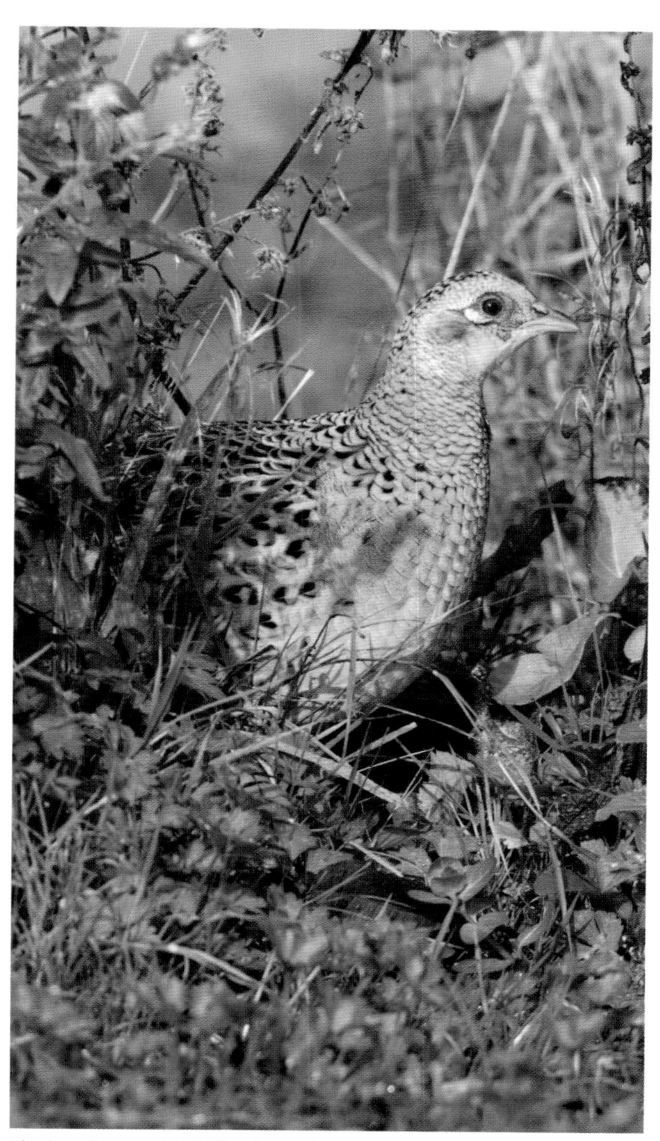

The hen Pheasant is dull-coloured.

GREY PARTRIDGE

Perdix perdix

In some areas this game bird will, along with the Common Pheasant, come and 'hoover up' seeds in rural gardens. The species will be familiar with food put out in countryside areas for game feed. Like the Common Pheasant and Red-legged Partridge it is bred and released extensively for shooting for sport; consequently, it is used to the people involved with rearing it and no longer as wary as it once was. Adults are rotund and erect-looking birds with a brown back, and a pale belly with a black horseshoe-shaped patch. The chest and flanks are grey with brown and black barring, and the face is orange-brown.

LENGTH 30cm.

HABITAT AND DISTRIBUTION Farmland and open habitats across Europe, but becoming increasingly rare.

DIET Although these are mainly vegetarian birds that eat a range of plant material, including buds, grass, seeds and grain, they do take a few insects. The chicks eat only insects for several days after hatching as they cannot digest seeds and similar items.

NESTING In the wild nests in the margins around cereal crops, and in deep grass cover under hedges and similar plants. The female can lay up to 20 eggs.

Partridges will search along the edges of fields and hedges for natural food.

Partridges will often search for grit on country roads.

RED-LEGGED PARTRIDGE

Alectoris rufa

Often referred to as French Partridge because Charles II introduced it in the UK as a game bird in the 1600s, this species has been bred for shooting ever since and now outnumbers the native Grey Partridge. It is a smart, rotund bird with a sandy-grey back and pinkish belly, and a warm pale grey breast blending into black-and-white speckles towards an all-black gorget. It has wide black-and-brown bars on the flanks, and red legs and bill.

LENGTH 34cm.

HABITAT AND DISTRIBUTION Arable farmland and dry lowlands. Elsewhere in Europe it occurs in France and Iberia.

Red-legged Partridges will come to gardens in the country or the edges of towns.

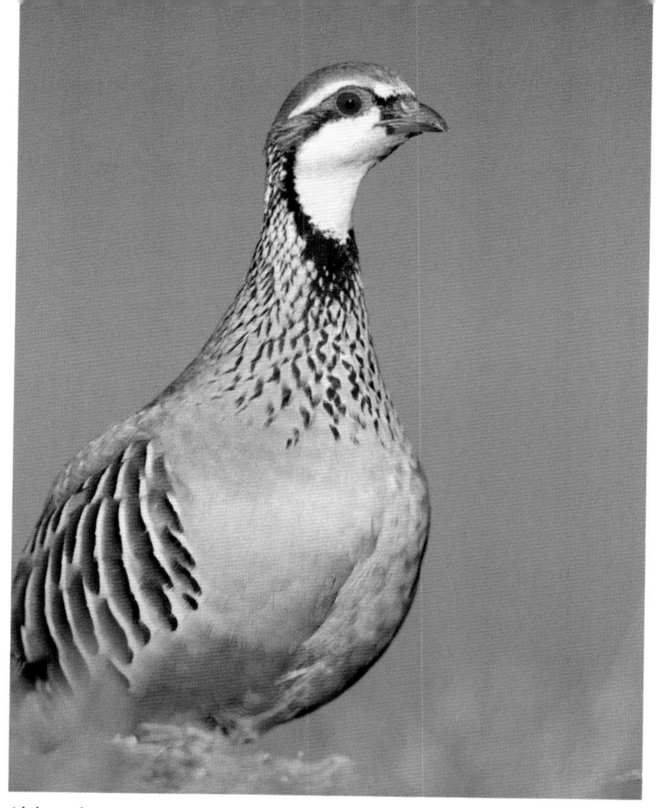

Although not native to Britain, the Red-legged Partridge is a very handsome bird.

DIET Mainly eats seeds and other plant matter. Due to the practice of game feeding in the countryside, Red-legged Partridges recognise bird-seed containers and will come to feeder sites in rural gardens, usually feeding on the ground. They are, however, able to get onto bird tables (as are Common Pheasants), where they can be greedy guests that feast on seeds.

NESTING The nest is built in cover and, very unusually, the female makes two nests. She lays a clutch of eggs in each, and the male and female brood and attend to feeding the chicks at the same time. The birds join up as a family party after the young leave the nests.

GREY HERON

Ardea cinerea

Tall with long legs and a long neck, the Grey Heron is a formidable bird that is mainly grey with black-and-white markings. In breeding plumage it has long black plumes on the head.

LENGTH 95cm.

HABITAT AND DISTRIBUTION Rivers, lakes and marshland are the main habitats, but Grey Herons also feed in coastal areas as well as on town and city park lakes and ornamental pools. They occur in the UK, across mainland Europe and in much of Africa, usually near water.

DIET Eats fish, amphibians, small birds, lizards, small rodents and even Rabbits. May come to gardens to raid ponds for goldfish and amphibians, usually doing so in the quiet hours of the morning before people are up and busy.

NESTING Nests in colonies, including in some city parks. The birds build huge nests, generally high in tall trees.

This heron is about to land. Usually when in flight the neck is tucked into a tight 'S' shape and looks quite short.

When not feeding, herons perch up high in trees.

RED KITE

Milvus milvus

The Red Kite is a large bird with a 1.5m wingspan and a deeply forked tail, which is very actively used as a flight control when it is in the air. Overall it is a brick red-brown colour, with a paler greyish head and yellow bare parts. Red Kites have a *wee-ooo-we-oo* display call that is high pitched and uttered in flight. Often several birds are seen together wheeling around in circles as they use thermals to gain height.

LENGTH 64cm.

HABITAT AND DISTRIBUTION Not so many years ago the Red Kite was a rarity in the UK, but Scandinavian birds have been introduced to some English counties, where they are doing well (genuinely British birds are restricted to Wales). The species is found in mainland Europe and western North Africa.

Red Kites swoop down to grab food with their strong feet and talons.

DIET Feeds on carrion, but also takes earthworms, small rodents and large insects. As the population has risen in recent years, the birds have increasingly been seen at refuse tips scavenging for food. In some areas people regularly feed Red Kites with meat scraps, so they are not an uncommon garden visitor. They are, however, more likely to be seen flying above gardens.

NESTING Nests in wooded areas with open spaces, building a scruffy nest of mud and twigs, often incorporating man-made material like plastic bags. The female lays 2–3 eggs and the male brings food to her. Both parents feed the young.

It is unusual to see Red Kites on the ground for long, unless the carrion they are feeding on is large.

SPARROWHAWK

Accipiter nisus

This is a stunning bird and a highly skilled hunter, scything its way along the backs of hedges and walls and popping up for the kill with well-practised precision. Males are small, not much bigger than a Mistle Thrush. They are a smart slate grey-blue above with whitish underparts covered in orange-brown barring, and have orange-brown cheeks. Females are bigger and grey-brown above with brown barring on the breast. The legs are yellow, as is the bill, which has a black tip. Sparrowhawks do not hover like kestrels; they employ either a slow flap-flap glide flight or a high-speed low-level flight.

LENGTH 28–38cm.

HABITAT AND DISTRIBUTION Open woodland, parks, gardens and hedgerows throughout Europe. In gardens it is most likely to be seen on garden fences or bathing in pools.

Sparrowhawks bathe regularly, often around the same time each day and usually in a favourite pool. This is a female.

The male is smaller and greyer than the female.

DIET Small birds are the main food. If you feed garden birds it is important to understand that by default you will be providing for predators as well – this is nature's way of honing a healthy population by removing the sick and weak, and the Sparrowhawk has its place in the way nature works. You can make it harder for Sparrowhawks by using caged feeders and placing upright canes around them to impede their access; moving the feeders around will also help to make small birds a bit safer.

NESTING Nests in trees on a flat platform of twigs. Sparrowhawks usually lay 3–7 eggs, which the female incubates while the male supplies her and the young when they hatch with food. Both adults feed the young as they get bigger.

COMMON BUZZARD

Buteo buteo

This is a large bird of prey with a wingspan of 114–128cm. It is generally brown overall with a yellow bill that has a black tip. Quite variable in colour and the way it is marked, it also tends to look rather 'scruffy'.

LENGTH 50–57cm.

HABITAT AND DISTRIBUTION Wide open, mixed countryside with stands of trees. The species is widespread in Europe, but in recent times was persecuted down to low numbers in the UK. Happily populations are now increasing in Britain.

DIET Hunts rodents and other small mammals. Buzzards have been moving into populated areas, where they scavenge discarded fast food, road kill and other waste, so while they do not strictly fit into the 'garden bird' mould, they are highly likely to be seen over and near gardens in many areas. If you have a large garden in the countryside you should be able to attract them by putting out bait of road-killed Rabbits and Pheasants.

Buzzards can often be seen on the ground hunting earthworms.

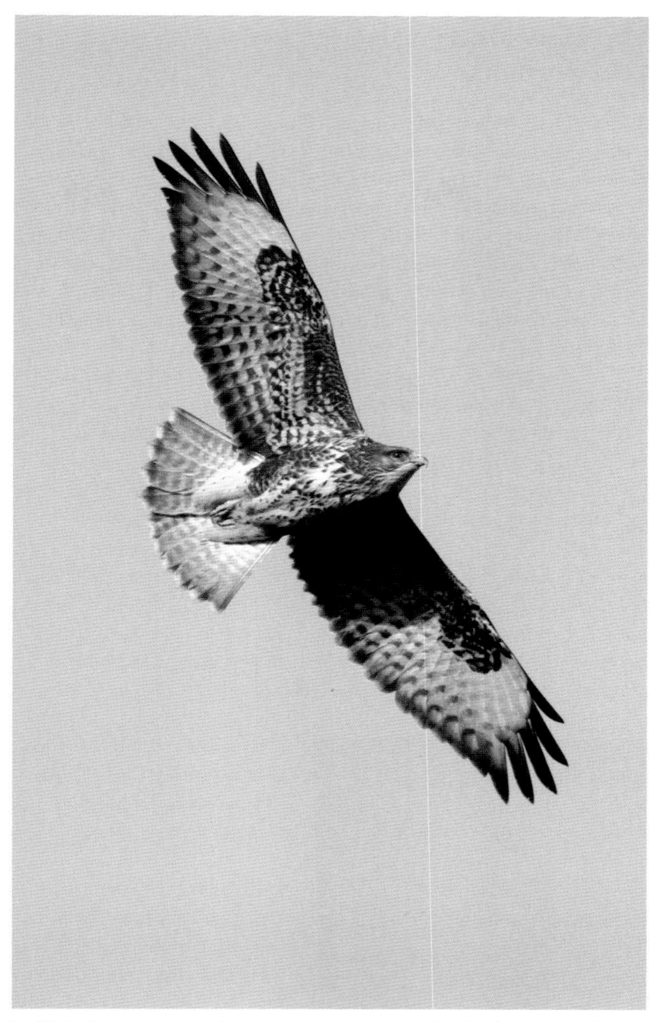

In flight buzzards can show varying amounts of pale plumage and some can look quite different.

NESTING Nests in a tall tree and lays 2–3 eggs. Once hatched the young are strong enough to fly after about eight weeks.

PEREGRINE FALCON

Falco peregrinus

A bird of wild places, towering cliffs and rock faces, hunting Rock Doves and other unwitting victims, the Peregrine Falcon power dives onto its prey in flight at speeds of 300km per hour. Until recently the population was low due to the use of organochlorine pesticides in agriculture and persecution by pigeon racers. Since the restriction of the pesticides, it has increased slightly and the birds have moved into the concrete towers and glass cliffs of cities, where they are doing well with a supply of Feral Pigeons to hunt.

The Peregrine is a large bird with a wingspan of around 1m. It is slate-grey in colour above, and pale below with dark spots and dark barring under the wing. It has a bold black moustachial stripe, and yellow legs, bill and eye-ring.

LENGTH 40–50cm.

HABITAT AND DISTRIBUTION Cliffs, mountains and urban areas throughout Europe. Peregrines are more likely to be seen in flight above a garden than in it. The highest chance of seeing one may be in winter, as northern birds move southwards; Scandinavian birds move to the UK at this time.

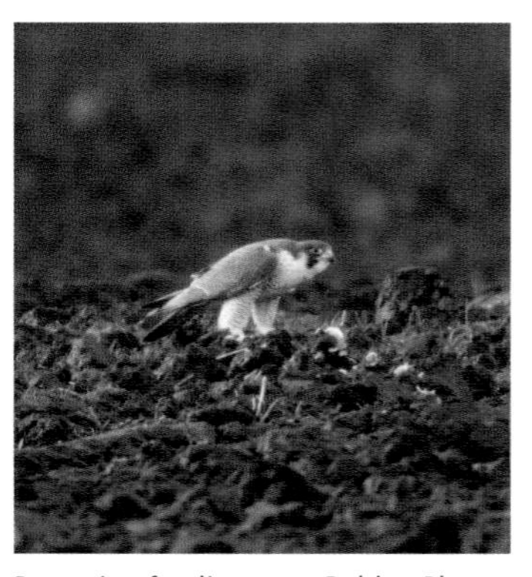

Peregrine feeding on a Golden Plover.

DIET Feeds on birds, with Feral Pigeons comprising some 80 per cent of its diet in some cities.

NESTING The nest is on a rocky outcrop, or (in cities) on a tall building, often in a nest box or tray put out for it. The 2–6 eggs are incubated by the hen; the young are fed by both adults.

Peregrines are supreme flying machines, perfectly adapted for catching birds in flight.

COMMON KESTREL

Falco tinnunculus

This common, widespread and abundant bird of prey can often be seen from a car or train, perched on signs, posts and street lamps. Kestrels have recovered well from the problems caused by the use of pesticides in the 1950s and '60s, when numbers fell drastically.

The Common Kestrel has a wingspan of 65–80cm, and its long wings and tail, and short neck, make for a distinct silhouette when it is flying. Its most obvious identification feature is also seen in flight: it hovers faced into the wind, matching the wind's speed and hanging in the air when it hunts for prey. The male is a rich brown colour with grey-black spots on the back; underneath it is a warm buff with brown streaks that are slightly closer together on the flanks. The head, tail and rump are grey. The female is slightly larger and similar, but she is brown on the head and rump, with a brown tail with dark bands; juveniles look similar.

 **LENGTH** 32–39cm.

This male, with a grey head and tail, is hovering while hunting for small rodents.

HABITAT AND DISTRIBUTION Farmland, moors and other open areas, and breeds in cities and towns across Europe. It is most likely to be seen flying above gardens, or visiting them to drink from or bathe in birdbaths and pools.

DIET Feeds on earthworms, small mammals and anything in between, including crickets, grasshoppers and beetles.

NESTING Nests in crevices in buildings, rocky outcrops, holes in trees and gaps between bales in straw stacks. Kestrels do not build a nest as such, but use old nests made by other species. The female usually lays 3–6 eggs. Both adults take care of brooding and feeding the young and looking after them for some time after leaving the nest until they have learned to hunt for themselves. Kestrels do use nest boxes in large gardens that are quiet and undisturbed, and within easy reach of grassland for hunting; in suitable places they may become resident.

The female kestrel is much browner in colour.

MOORHEN

Gallinula chloropus

Members of the rail family, Moorhens look black but the mantle and wings are actually dark brown. There is a red shield and base to the bill, which changes to yellow at the tip, and the legs are also yellow with red garters above the knee joints. The underside and outer edges of the tail are white – the bird flicks its tail as it swims and walks, so the white tail is a very obvious field marking. Moorhen numbers have suffered as a result of predation by American Mink *Mustela vison* since this mammal was released into the wild. However, the population is still high enough for the birds to be seen on most stretches of water. Their distinct *churruck* call is loud and diagnostic.

LENGTH 30cm.

HABITAT AND DISTRIBUTION Common and widespread throughout Europe, Asia and into India, and also South Africa. A common and tame bird in town parks, Moorhens have become used to being fed by people. They are also regular visitors to gardens in the right locations.

Moorhens have big feet to enable them to walk on floating vegetation.

DIET Eats aquatic insects and vegetation, snails and insects. In gardens Moorhens also eat most commercially available bird foods, and pet food if pets are fed outside.

NESTING Breeding starts in March with the nest being built on a bough or other object near or overhanging water. The male gathers twigs and man-made materials, with string and crisp bags being used in town parks. The female puts together a loose platform with some finer twigs and reeds for the bedding, which the speckled glossy grey-green eggs are laid on.

The young are black with pink bald heads, and are able to swim a couple of days after hatching.

Moorhens often stand on one leg then change to the other, probably to conserve body heat.

COOT

Fulica atra

A very easy bird to identify, the Coot looks like a small black duck. Seen out of the water the pale blue-grey feet with lobed toes (slightly reminiscent of Horse Chestnut leaves) are very different from the webbed feet of ducks. This feature, combined with the shape of the white beak and facial shield, is typical of rails. The Coot is not secretive like many of its relatives. In winter, birds flock in large groups and are noisy, making a variety of cackling, clucking and sharp calls at all times of the day and night. They are reluctant to fly, usually preferring to half run and flap across the water's surface, either to escape a threat or to fight each other when competing for females during breeding.

LENGTH 38cm.

HABITAT AND DISTRIBUTION A bird of town park lakes, ponds and pretty much any area of still and slow-moving water. Common and widespread, generally favouring larger and more open water bodies than the Moorhen.

Territorial fighting is common.

Coots spend a lot of time swimming and feeding.

DIET Broad diet of buds, vegetation, insects, fruits, seeds and algae. Coots often dive head-first to reach underwater food items; they also eat the eggs of other birds. They may visit gardens with ponds or near water, mainly to feed on the ground, although they can learn to copy other species and also use bird tables.

NESTING The nest is usually on vegetation around water, built so that it floats up and down, or placed above the water on a solid branch. It is constructed mainly from twigs and reeds, but in town parks all sorts of litter, such as string, polystyrene and crisp bags, may be incorporated. Moorhens can lay up to 10 eggs in each of 2–3 broods every year. The young have a high mortality rate due to predation by Grey Herons, rats, gulls and even their own parents.

LAPWING
Vanellus vanellus

The Lapwing, also called Peewit or Green Plover, is one of the most attractive wading birds of the region. It may seem strange to include it as a garden bird, but in quite a few places where the habitat is right, such as the Hebridean Islands and Yorkshire Dales, Lapwings feed on lawns and even nest on them.

The Lapwing is mainly black and white; the green is on the back, which has a metallic sheen with shades of purple visible in some lighting. The male has a long crest extending from his black crown, black facial markings and a black breast. Females are similarly marked but have shorter head plumes, less obvious facial markings and chocolate-brown flecks in the black of the crown. Unusually among waders, Lapwings (and their close relatives) have round wings instead of more pointed wing-tips; the wingspan is 67–87cm. The 'out-of-control', tumbling display flight on the birds' breeding grounds is performed by the male, which makes a wheezing yet sharp *pee-whit-wee-wit-ee-it* call; shorter calls of just *pee-whit* are the main contact notes.

 LENGTH 28–33cm.

The Lapwing looks black and white from a distance but has iridescent green and purple in its coloration.

HABITAT AND DISTRIBUTION Widespread but decreasing breeding bird on farmland, low moorland and edges of large lakes.

DIET Food includes insects, worms, molluscs and some vegetable matter. In gardens in suitable areas, Lapwings come to mealworms and water if they are put out for them. Water is especially appreciated when the weather is dry.

NESTING Lapwings nest on the ground on cultivated land, and in short grass on marshes or moorland. They usually choose a slight ridge or hump with a good view of any likely predators, which are warned off with noisy attacks – they will even threaten sheep and horses that come too near the nest or chicks.

Lapwing in flight.

FERAL PIGEON

Columba livia var. *domestica*

Descended from the now rare Rock Dove *Columbia livia*, which is naturally found breeding on coastal cliffs, in towns and cities Feral Pigeons use urban buildings as their nesting 'cliffs' and make a good living from scraps, discarded fast food and food given to them by people. They typically have black wing-bars and a white rump, although interbreeding with homing pigeons has resulted in a wide variety of plumage colours, from white, through very dark grey to pale fawn.

LENGTH 33cm.

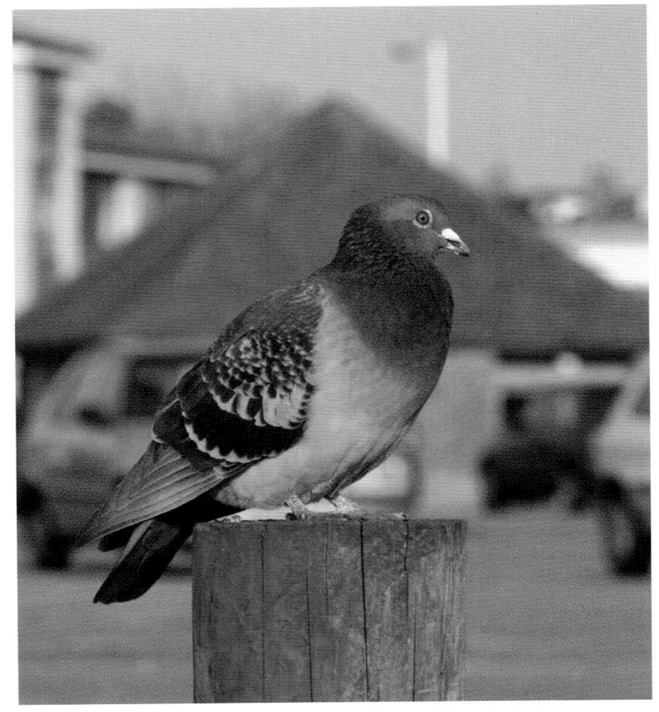

Feral Pigeons can vary in colour quite a lot.

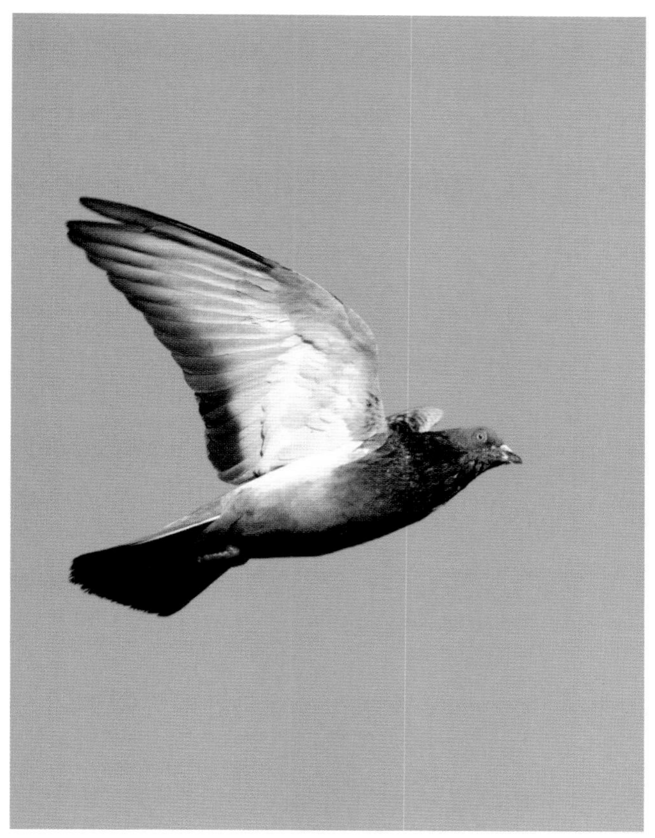

Feral Pigeons are a familiar sight in almost every town and city.

HABITAT AND DISTRIBUTION The common pigeon of Europe's towns and cities.

DIET Feeds on seeds, grains and discarded human food.

NESTING Feral Pigeons are able to breed all year round because of food abundance coupled with relatively high ambient temperature in towns. They seek out nest sites that are warm and out of the wind, such as hollows and crevices in buildings.

STOCK DOVE

Columba oenas

The Stock Dove is a wary bird, although it is also sociable and gregarious and is often found with Collared Doves and Woodpigeons. It is a small pigeon with a blue-grey back, a lilac-pink breast and an iridescent green patch on either side of the neck. Its eyes are black, and there is a black band on the end of the tail and two similar-coloured wing-bars; the legs are pink.

LENGTH 30cm.

HABITAT AND DISTRIBUTION Woods, farmland, parks and gardens across Europe into Asia. Half the world's population is found in the UK, and it is the rarest of the pigeons in the region. Populations declined in the 1950s and '60s due in part to the use of organochlorine seed dressings, but they are now on the increase.

DIET Feeds mainly on seeds, grains and fruits, and some leaf and flower buds. Stock Doves do come to bird tables and seeds on the ground, and are most likely to be found in quiet gardens with little disturbance.

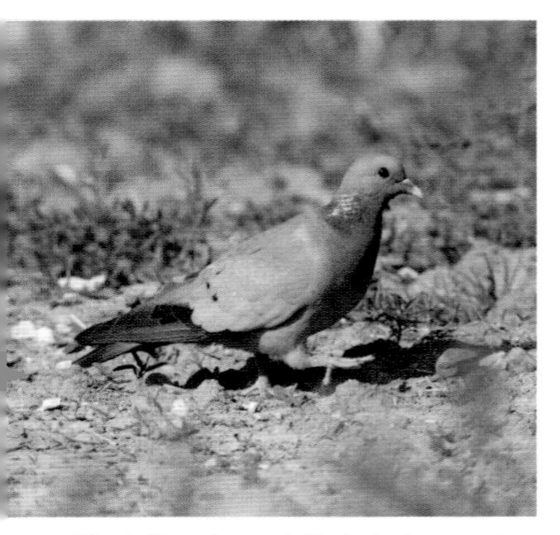

Stcok Dove is most likely to be seen in rural gardens.

NESTING Nests in holes in trees and buildings, and in the gaps between stacked straw bales. The nest is made up of twigs and dry leaves. Both adults share the incubation and feeding duties. The clutch is just 1–2 eggs at a time, with up to three broods a year.

The black, bead-like eye is an obvious identification feature.

WOODPIGEON

Columba palumbus

This is a big bird with an uncanny knack of looking like several other species when perched on a post or wire; depending on its posture it can be mistaken for a Kestrel or owl. It is generally grey with a pinkish breast and a white patch with an iridescent purple patch on the neck. It has a yellow eye with a black patch on the iris, making the pupil look an odd shape, and a pinkish bill and legs. On the ground Woodpigeons appear overweight because of the way they waddle. However, in flight they are swift and direct, and their bold white tail and wing markings show prominently.

Woodpigeons have a *ru-hoo ru roo-hoo* call that is repeated several times in succession. During display flights the male flies up, then claps his wings together with a sharp crack before gliding down with wings and tail spread out.

LENGTH 40–42cm.

HABITAT AND DISTRIBUTION Woodland, parks, gardens and farmland across Europe.

DIET A seed and grain eater, but takes bread and cereals at feeding sites. Woodpigeons often become tame and hog the food, putting off other, smaller birds. Because most of their food is dry they drink a lot. Unlike other birds they do not scoop up the water and tip the head back to swallow it, but put the bill on the water and either suck or use the tongue to pump water in.

NESTING The nest platform is a flimsy-looking twig structure, which is built by both parent birds. Two glossy white eggs are normally laid. Both adults incubate them and feed the young.

Woodpigeon is the biggest member of the family.

Over the past 30 years, Woodpigeons have become frequent and bold garden visitors.

COLLARED DOVE
Streptopelia decaocto

It is hard to believe that before breeding for the first time in the UK in 1956 this bird was a rarity. It originated in Asia, and from the end of the 19th century spread north and west, possibly as a result of the increasing use of intensive and mechanised farming.

The Collared Dove is typically pigeon-shaped, but smaller and daintier compared with many of its relatives. Both sexes are an overall grey-buff in colour, turning to paler pinkish-buff underneath. The eye is red with a white ring around it, the bill is black and the birds have a black half-collar outlined with white – the feature that gives the species its common name. The call is a *coo coo coo* repeatedly made from quite early in spring and mistaken by some for a Cuckoo's call. The birds are almost always seen in pairs and seem to be very loyal to their partners.

 LENGTH 32cm.

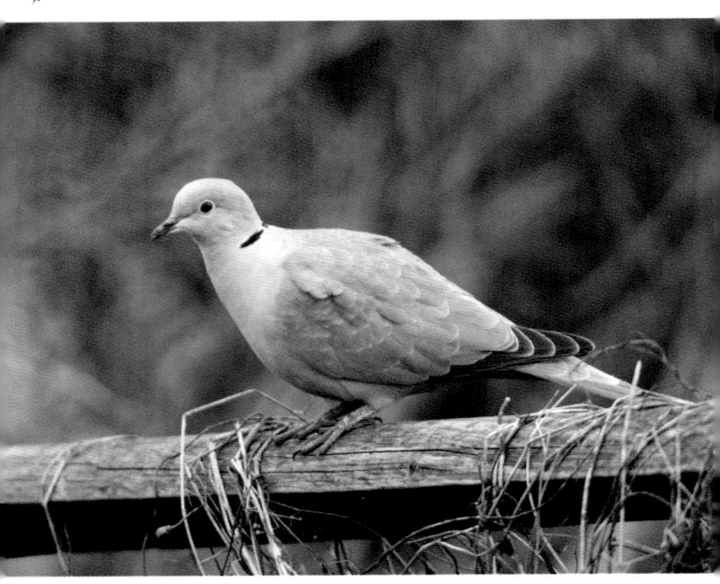

Juvenile birds often lack the black collar marking.

HABITAT AND DISTRIBUTION Towns, gardens and farmland with hedges across Europe. Collared Doves are now common in parks and gardens. Their success is due to their ability to live in close proximity to humans. Very few nests occur in open countryside away from built-up areas.

DIET Eats almost any food put out for it, including grain, insects and breadcrumbs. The birds are gregarious and form flocks with high numbers, which gather around grain stores especially.

NESTING The nest is a flimsy structure of twigs in which the female lays two white eggs. Unusually, the female incubates at night and the male does so during the day. The species is a prolific breeder near abundant food, with 3–4 broods a year not being uncommon.

Collared Doves perch in the open and are easy prey for Sparrowhawks.

RING-NECKED PARAKEET

Psittacula krameri

Also called Rose-ringed Parakeet, this species first bred in the UK in 1969, a long way from its Asian homeland. It is an adaptable and successful bird; the climate in the UK is not disimilar from that of the higher altitude climate in its homelands, so it is not surprising that it is doing well. The birds are unmistakable, being bright emerald green in colour with a long tail. The male has a black-and-pink ring around his neck, and the adults' legs are greenish-grey in colour.

LENGTH 40cm.

HABITAT AND DISTRIBUTION In Europe mainly parks with old trees. The UK population is increasing and there are several roosts of 5,000 or more birds around south-west London and the Heathrow area; they have been seen as far north as Sheffield. The population probably originated from escapes from aviaries.

Despite the bright green colour they can be difficult to see, even when you can hear them.

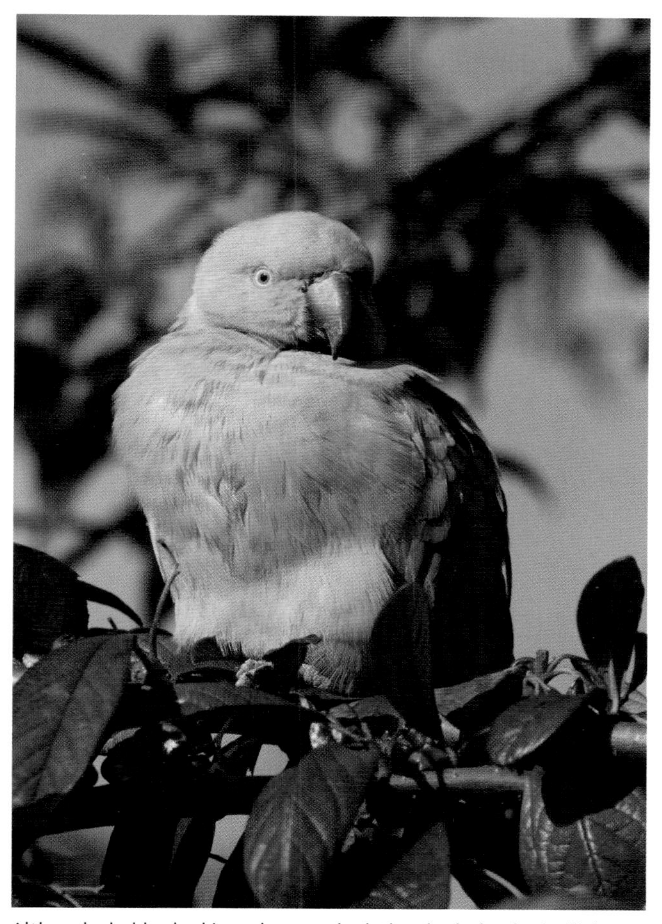

Although chubby-looking when perched, they look slender in flight.

DIET Feeds on buds, fruits including berries, nuts, seeds and vegetables. Ring-necked Parakeets are regular at bird tables in the areas they have colonised.

NESTING The nest is usually in a hole in a tree and the female lays and incubates 3–7 white eggs.

CUCKOO

Cuculus canorus

The bird that heralds the spring with its familiar call, the Cuckoo is sadly in decline. The fall in its numbers is so alarming that it is on the Red List for requiring special conservation attention. It is a largish, grey-backed bird with a grey head and throat. It is boldly dark barred on its white lower breast and belly, and has yellow legs, a yellow bill with a black tip and red eyes. Some females may be rufous. With a wingspan of 55–60 cm, it is about the size of a pigeon and resembles a Sparrowhawk with its 'bird-of-prey' shape in flight.

LENGTH 33cm.

HABITAT AND DISTRIBUTION Habitats include moorland, heaths, open woodland, parks and large gardens. Summer visitor to Europe, arriving in spring and leaving in late summer.

DIET Feeds on insects.

Most Cuckoos are grey in colour.

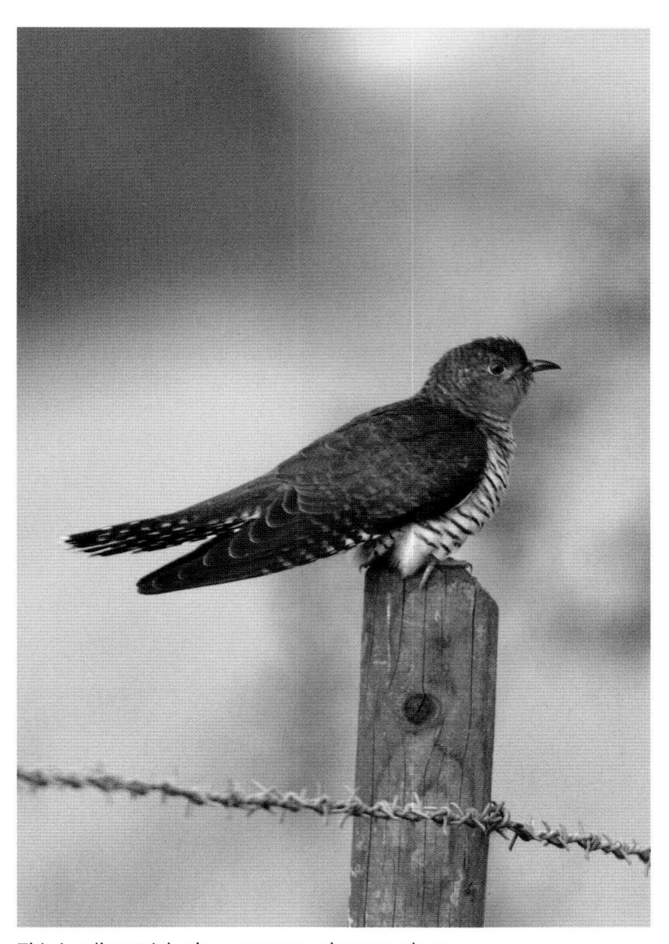

This is a 'hepatic', a less common plumage phase.

NESTING The Cuckoo is well known for its habit of laying eggs in other birds' nests. It is a brood parasite mainly of warblers and pipits. While it is not commonly found in gardens, it does lay eggs in the nests of Robins, Dunnocks and wagtails, which is why it is included here. Young Cuckoos have the ability to mimic the begging calls of the chicks of the birds that are raising them.

BARN OWL

Tyto alba

Barn Owls will not be seen at a bird table, but they are present around rural gardens, tucked away roosting in buildings and holes in trees. The heart-shaped face and big eyes give them an endearing look. The birds have predominantly white underparts and face, with a golden-buff mantle, wings and head, and yellow feet with large, sharp, pointed talons. Males are usually paler than females, but both sexes are similar in size, although females sometimes appear larger. Females also have banding on the tail that is most obvious when they are flying, although it can be difficult to tell the sexes apart. Both sexes – but particularly females – can have varying amounts of dark spots on the flanks.

Barn Owls are not very vocal, making soft hissing and beak-clicking noises as contact calls to each other, but these are not often heard. The old name for the species was Screech Owl because of a rasping screech call that is made infrequently. The birds are generally nocturnal but do hunt in daylight when they have young or when the weather prevents hunting at night.

LENGTH 36–40cm.

Barn Owls 'mantle' their prey to avoid it being stolen by Kestrels.

HABITAT AND DISTRIBUTION Fields, meadows and marshes that have plenty of open country with rough grassland for hunting. Barn Owls can be found in most areas of the UK, but are scarce in northern Scotland.

DIET Feeds mainly on rodents, mostly rats and voles.

NESTING Barn Owls pair up in January and lay eggs in February. They often nest in rural buildings such as barns and farm outbuildings. The eggs take about 30–34 days to hatch, and hatching is timed to coincide with small mammals' new crops of youngsters, which serve as food for the chicks. Barn Owls readily use nest boxes when they breed, and putting up a suitable nest box in a large rural garden increases the chances of having them in a garden. They sometimes roost in nest boxes outside the breeding months, so may be present all year round. As a Schedule 1 protected species they must not be disturbed.

Between January and April, when they have growing young to feed, Barn Owls can often be seen out in daylight.

LITTLE OWL

Athene noctua

This introduced bird is now quite a common sight in the British countryside, often seen perched on a bough, close to the trunk, snoozing in the warm sun. A small owl, it can look quite rotund when the feathers are puffed out to keep it warm. It is brown with white speckles and similar below but more streaked. The head is large with piercing yellow eyes set under obvious eyebrow markings.

LENGTH 22cm.

HABITAT AND DISTRIBUTION Open country with farmland and scattered trees, and open woodland. Found right across Europe except the north and north-west, into Asia and down into Africa, with 13 races across its range, the Little Owl was introduced in the UK in around 1842 by Thomas Powys at Lilford Hall, Northamptonshire.

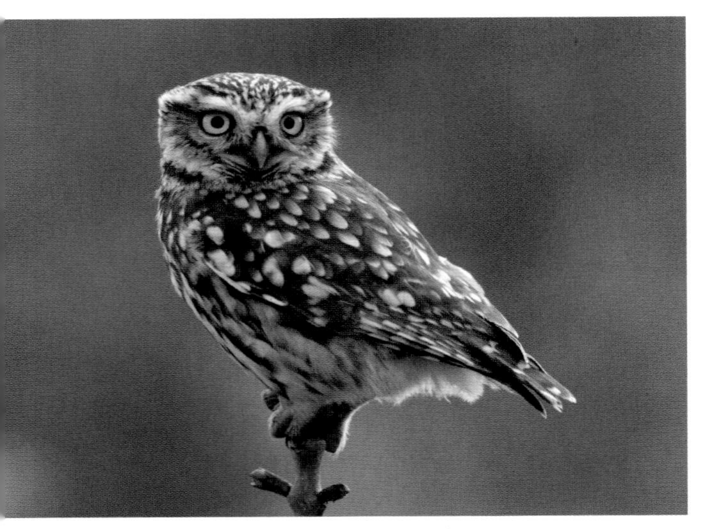

Adult Little Owl on a favourite hunting perch.

DIET Eats insects, worms, small amphibians and small mammals. Often hunts in country lanes, gathering wind-blown insects that are caught up in the overhanging grass. Crane-flies seem to be collected in this way especially during breeding, usually for feeding chicks. Little Owls can become quite tame in places where they live close to human habitation. They may come to earthworms if they are placed in a regular spot; otherwise they do not visit bird tables.

NESTING Nests in holes in trees. It may inhabit large gardens containing trees with holes or old buildings that provide suitable nest sites.

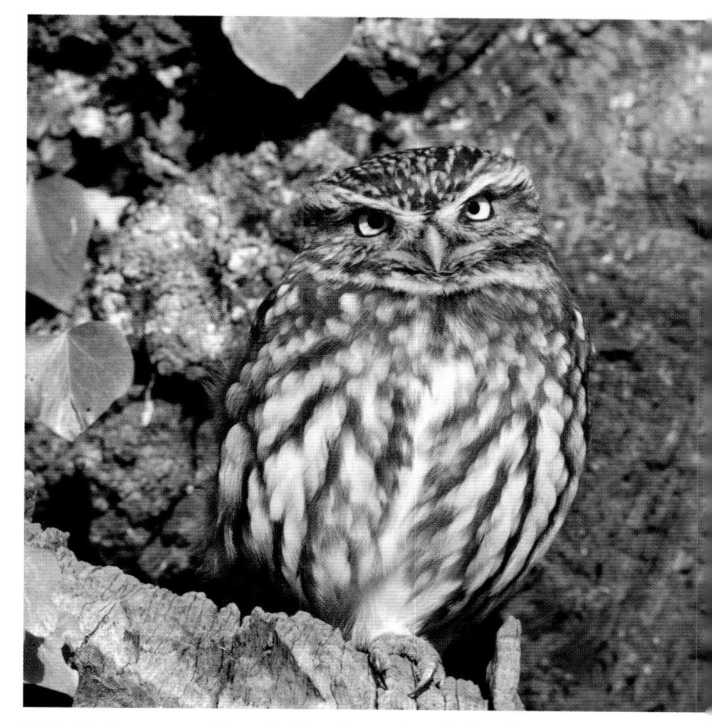

Little Owls can sometimes be found snoozing during the day.

TAWNY OWL

Strix aluco

A plump-looking brown owl that is mainly nocturnal, unlike other owls in the region the Tawny Owl does not hunt on the wing. It prefers to perch and watch the ground below for prey. The only signs of Tawny Owls in your garden might be the sounds of small birds making repeated alarm calls to alert other birds of the owl's presence, or the calls of the owls at night. The familiar *twit-twoo* is in fact two birds making contact calls to each other; more accurately the sound of the first bird is a sharp *kee-wick*, and the return call of the second bird is *hoo-oo-ooo*.

Tawny Owls usually hunt just after dusk. Once they have caught enough prey they roost until it is digested, regurgitate the pellets, then hunt some more, with about three hours between hunting sessions. On wet nights they hunt from overhanging branches near roads, often on the leeward side of woodland, usually taking many small prey items like worms and beetles. On these nights they are easier to see, possibly because they hunt for more time than they spend roosting.

 LENGTH 38cm.

Tawny Owl chick perched outside a nest cavity in a broken tree.

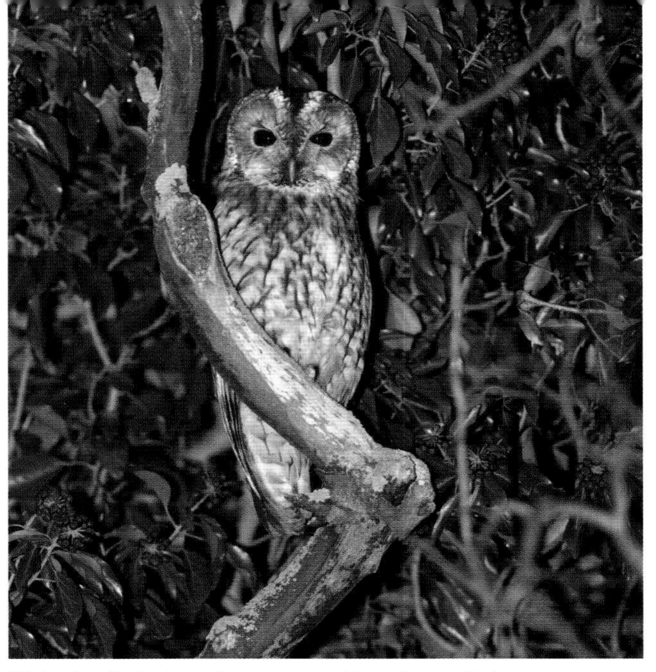

Night time is the best time to see and hear Tawny Owls.

HABITAT AND DISTRIBUTION Deciduous woodland, forests and parks with mature trees. Widespread throughout Europe, except the north and Ireland, and Asia into the Kashmiri region and down into North Africa, with eight recognised subspecies.

DIET Eats anything from insects and earthworms to small rodents, with frogs and toads in between. In urban areas the diet has a higher percentage of small birds in it, and some owls have even been known to catch fish from shallow pools.

NESTING Nests in holes in trees and buildings, and sometimes in squirrel dreys and old Magpie nests. Breeding starts in March and the female lays 2–7 glossy white eggs. She incubates them for about four weeks, and when they hatch both parents feed the young. Tawny Owls may nest in specifically designed nest boxes in large rural gardens. The birds are sensitive to disturbance and may be aggressive while nesting, when they are best left alone; gardens with pets and children are therefore inappropriate.

LONG-EARED OWL

Asio otus

This medium-sized owl is easily confused with the Short-eared Owl at a distance, but when seen at close range the rich orange eye colour and the more obvious ear-tufts are key identification features; Short-eared Owls have a yellow eye. In flight Long-eareds can be seen to have shorter, more rounded wings, similar in shape to those of Tawny Owls, while Short-eareds have longer, slimmer wings.

The Long-eared Owl is a migratory species and may turn up in gardens when on passage, usually roosting in deep cover by day, and often in small groups – sometimes one or two birds will be right out in the open.

LENGTH 31–40cm.

HABITAT AND DISTRIBUTION Occurs across Europe into Asia, and also in North America and Canada, usually preferring open countryside and evergreen woodland. In a suitable area you may be lucky to have this bird roosting, but otherwise it is an infrequent garden bird.

DIET Hunts small rodents and birds.

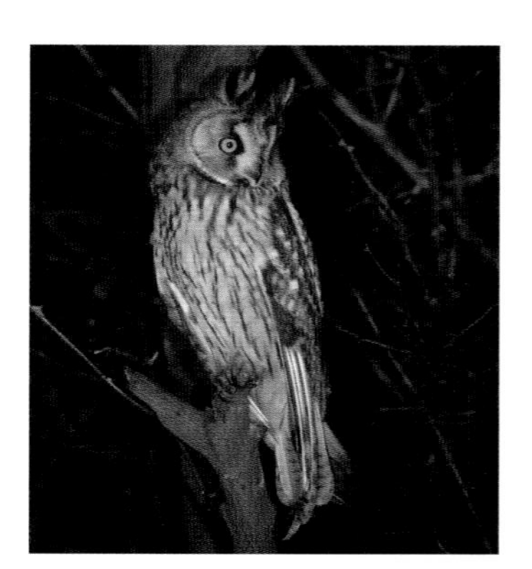

The Long-eared Owl is not an easy bird to see, but they do sit out occasionally.

NESTING Nesting is in January–July, with a pair sometimes choosing the old nests of other birds such as crows. They may also nest in artificial baskets. Usually 4–6 eggs are laid and they hatch in about 27–30 days.

When roosting, Long-eared Owls are able to adopt a slim, erect posture in order to help camouflage themselves.

SHORT-EARED OWL

Asio flammeus

This is another bird that only fits in as a garden bird in some locations because of its habitat requirements. It is a diurnal hunter that can be seen at any time of the day. It is similar in appearance to Long-eared and Tawny Owls, but has noticeably longer and slimmer wings (95–105cm wingspan), a characteristic visible in flight. The ear-tufts are not always visible because they lie flat. The plumage is buff to gingery-buff with dark blotches, and the tail has four distinct bars on it. The face has variable amounts of black – some individuals can be quite black-faced, which makes their bright yellow eyes stand out and gives the birds an 'angry' expression.

LENGTH 35–42 cm.

HABITAT AND DISTRIBUTION Migratory with the seasons in both hemispheres; in the north there is a big movement south as winter approaches. The owls often fly long distances across the sea, and the ones that winter in the UK come down from Scandinavia. Although this is not a bird that will occur at garden feeders, it may be flushed from roosting in the grass and scrubby areas of a big garden.

In winter the population increases as birds move south.

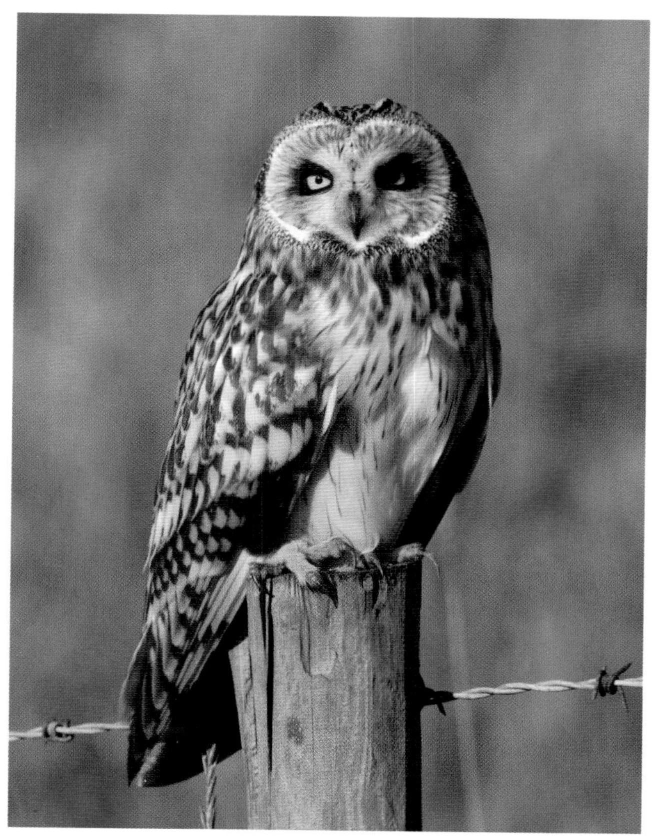

Fence posts are used for resting and as vantage points.

DIET Feeds on small mammals, particularly voles, and also some birds.

NESTING Nests and roosts on the ground in moorland habitat; elsewhere roosts low in bushes and scrub. Lays a single clutch annually of 4–8 smooth white eggs that take about 26 days to hatch. The female incubates and feeds the young. The male hunts for food and presents it to the female, which then passes it to the chicks.

COMMON SWIFT

Apus apus

The Common Swift is a summer visitor to the UK and mainland Europe. It is not a garden bird as such, but can be observed flying above gardens and may nest in the roofs of buildings nearby. It has a wingspan of 43–48cm and is quite a large bird. It is dark sooty brown with a white throat-patch. While the bill looks small it is only part of a large mouth surrounded by bristly hairs, perfectly adapted for catching insects in flight. The tail is of medium length and forked; it is usually closed rather than 'fanned', so that the fork is not visible, unlike in a swallow's tail.

On summer evenings Common Swifts make a screeching call as they fly low and fast.

HABITAT AND DISTRIBUTION Breeds in towns and villages. Summer visitor to northern Europe, except Iceland. Swifts have been in decline, probably due to a lack of suitable nest sites as old buildings make way for new ones.

DIET Feeds on insects, which it takes in flight, often several kilometres from its nest sites.

NESTING Swift nests are built in holes and under eaves. They are constructed from light grass, feathers, hair and leaves, all collected while flying and stuck together with saliva. Usually 3–4 matt white eggs are laid. Both parents take turns to sit for about three weeks until the young hatch, and both feed them. Swifts may use specially designed nest boxes.

KINGFISHER

Alcedo atthis

Often one's view of a Kingfisher is a brief glimpse of a two-tone missile darting along low over water, making a shrill *tsee-tsee* call as it disappears. The species is included here because it does show up in gardens that are close to water or have a suitable pond – as its name suggests, it eats fish.

The Kingfisher has a blue head with orange marks in front of and behind the eyes, and a white mark on the ear-coverts. The wings are greenish-blue and there is a bright blue streak from the neck, down the back and onto the tail. The brightness of the bird's colour is most apparent when the bird is in flight. The sexes are almost identical, except for an orange colouration with a black tip on the female's lower mandible. Young Kingfishers are similar to adults, but they have duller and greener upperparts, paler underparts, a black bill and, initially, black legs. The bill is also shorter with a fingernail-coloured growth patch at the tip. In the adult the dagger-like bill is about 4cm long and is used in hunting, which usually consists of diving head-first from a perch about 1.5m above the water.

LENGTH 17–19cm.

HABITAT AND DISTRIBUTION Rivers, streams and lakes.

DIET While the diet consists mainly of minnows, sticklebacks and other small fish, Kingfishers also catch small newts, frogs, toads and tadpoles, as well as dragonflies, damselflies and the larvae of several other insects. They may visit garden ponds to take small fish.

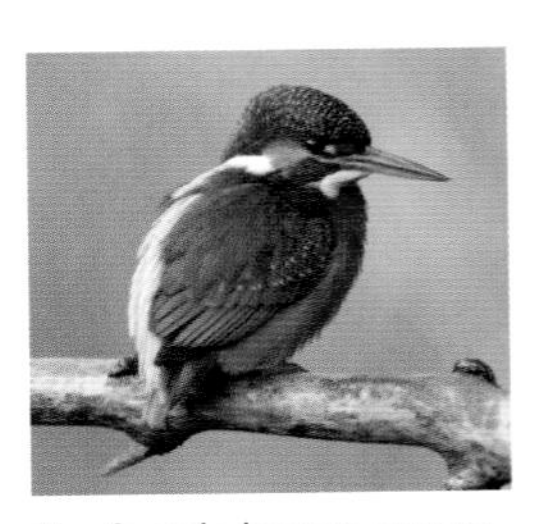

The female has an orange-coloured lower mandible.

NESTING Kingfishers nest in a 1m-long burrow dug in a sandy bank. They are a Schedule 1 protected species and must not be disturbed when breeding.

If you have a pond or brook in your garden, try placing a suitable perch for Kingfishers to use.

HOOPOE

Upupa epops

Although it is rare in the UK, the unmistakable Hoopoe is a widespread bird across Europe and Asia, and down into Africa. It is a spectacular bird, with a peachy brown-pink head, neck, breast and belly, black-and-white wings and a large crest that protrudes from the back of the head as far as the sharp-tapered bill. The wings are large and rounded, and have a distinct black-and-white pattern; during flight they are closed intermittently, resulting in bounding undulations that give the impression of a big butterfly. The bird's call is typically a trisyllabic *oop-oop-oop* sound, hence the onomatopoeic name.

LENGTH 26–32cm.

HABITAT AND DISTRIBUTION Although the Hoopoe is in decline in northern Europe, it has adapted to human activities and does well in olive groves, vineyards and orchards, particularly around the Mediterranean region. Due to its habitat preferences of lightly vegetated ground with rock faces, walls or trees with suitable cavities for nesting, it is generally only seen in suitable locations.

DIET These birds should be encouraged because their diet is insects, including many of the pest groups such as grasshoppers, locusts, beetles, earwigs and cicadas. They use the long, downcurved bill to extract insects from holes in the ground and rotting wood. Hoopoes also take larger prey like frogs and lizards; they beat much of their prey against a stone or branch to kill it, removing indigestible parts before eating. If the birds show up in a garden, it will usually be to search the lawn for grubs. They also use birdbaths, and water is likely to be the main attraction for them in a garden.

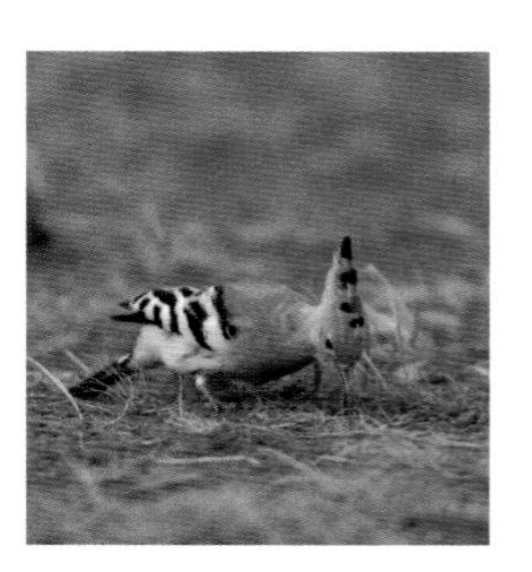

NESTING Hoopoes are territorial and pair bond for the nesting season. The male is aggressive and fights between rivals are frequent. A cavity nest site is chosen by the female (Hoopoes occasionally also use nest boxes), and she lines it with grass and moss and incubates the eggs. The male feeds her during incubation and helps to feed the young when they hatch. Usually the nest is messy and smelly; the female produces a foul-smelling secretion as an anti-predator strategy, and the young produce this substance as well. They can squirt it very accurately at anything they view as a threat, an ability they lose after leaving the nest.

The Hoopoe uses its long bill to probe for insects in the ground.

GREEN WOODPECKER

Picus viridis

A big bird with a 45–51cm wingspan, the Green Woodpecker is unmistakable, with a bright green back and wings, paler yellow-green underparts and a yellow rump that shows as a long yellow triangle when the bird flies away from you. The head has red on the crown and black moustachial markings; these have red patches in them on the males and are the only feature that can be used to tell the sexes apart.

The old country name for the species is Yaffle, an onomatopoeic interpretation of the bird's loud call, usually made as it flies off. It is a wary bird and difficult to get close views of. That said you are likely to see it feeding on lawns. It has an extremely long, sticky tongue that retracts into the back of its head, an adaptation that enables it to find food in deep holes and crevices. It also dust bathes on ants' nests for feather care – the ants attack the bird and squirt formic acid into its plumage, and it is assumed that this may help kill feather parasites.

LENGTH 30–35cm.

HABITAT AND DISTRIBUTION Deciduous and mixed forest edges, woodland, farmland, parks and large gardens.

Green Woodpeckers can blend in very well; you may only see them as they fly off.

DIET Feeds on grubs and hunts for ants and ant eggs, mostly feeding on the ground and taking a variety of invertebrate prey. Where it occurs in gardens, it may feed on fat cake, suet and mealworms. It is more likely, however, that it will visit for water; after feeding on ants it may come to bathe.

NESTING Nests in holes in trees. Despite its fearsomely lethal look the bill is relatively weak, so Green Woodpeckers usually choose to excavate holes in soft or rotting wood. The hole may take the male up to a month to dig. Once it is ready the female lays as many as six eggs, and both birds share the duties of incubation and feeding. The young emerge about 45 days after the eggs have been laid, and there is usually just one brood a year.

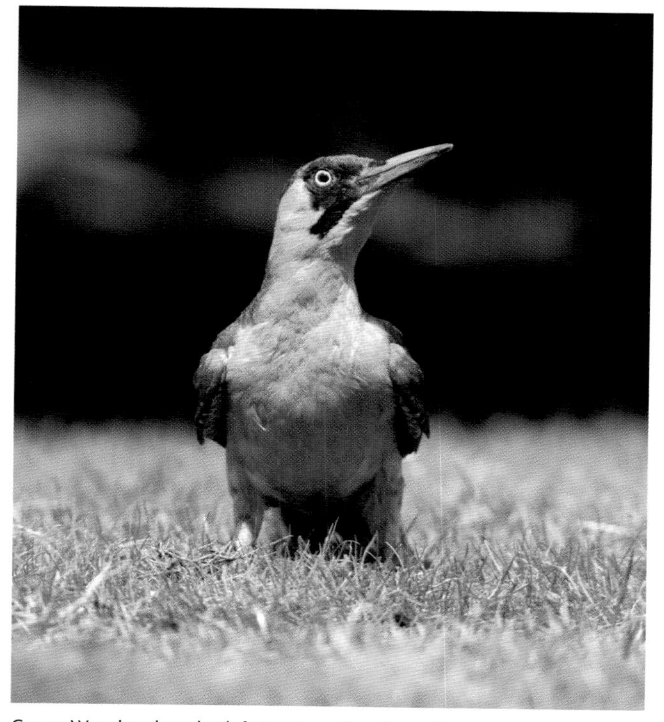

Green Woodpeckers look for ants and grubs on lawns.

GREAT SPOTTED WOODPECKER

Dendrocopos major

This is an unmistakable bird in its black-and-white garb. It is black on the wings and head, with a white breast and upper belly, white cheeks, neck and shoulder-patches, and white barring on the flight and tail feathers; the abdomen, vent and undertail-coverts are crimson. The sexes differ in that the male has a vivid red nape, while the female does not. Juveniles look similar and have a red patch on the forehead that extends onto the head. The birds are usually seen clinging to the trunks of trees using the stiff tail as a support, although they do also perch across branches. The similar Lesser Spotted Woodpecker *Dendrocopos minor* is a smaller bird, about the size of a House Sparrow, with fine white barring on the back.

LENGTH 24cm.

HABITAT AND DISTRIBUTION Woodland, large gardens and parks in the UK and right across mainland Europe to Asia. The species is very common and on the increase. It has recently started to recolonise Ireland, where it had been extinct for about two centuries. The Lesser Spotted Woodpecker inhabits deciduous forests, often near wetlands. In the UK it occurs in southern England and Wales, and is declining.

A male bird, with the red patch on the nape of its neck.

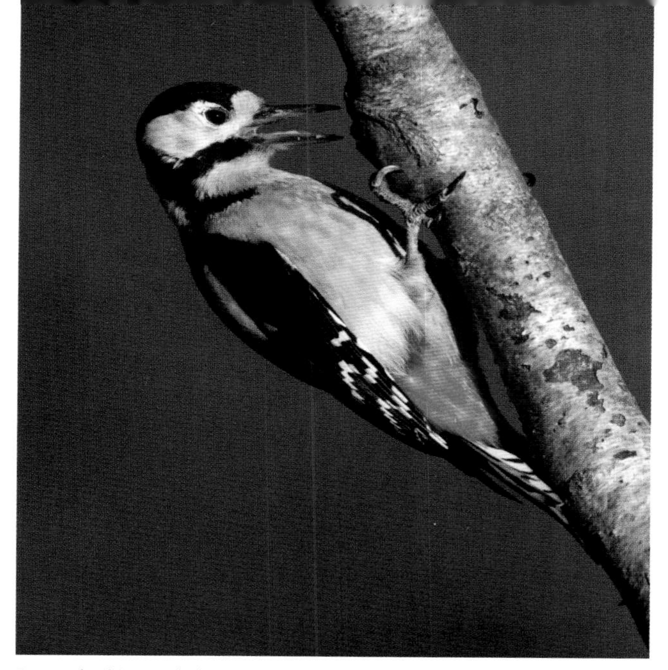

Fat pushed into a hole in a branch has attracted this female.

DIET The diet consists of a wide range of bugs and larvae, and the sharp bill and long, sticky tongue are the tools the birds use for extracting these from behind bark and crevices. They also feed on fruits including berries, as well as on the chicks and eggs of smaller birds. As regular birdfeeder visitors there is little that they will not eat. Peanuts seem to be a favourite, as well as suet cake. They also eat live or dried mealworms, wax worms, and most seeds but especially sunflower kernels. Woodpeckers can damage flimsier plastic and wire feeders by pecking through them to get at the food. The Lesser Spotted Woodpecker feeds mainly on insects, particularly grubs and beetles. It rarely visits bird tables, although it may take seeds in gardens.

NESTING Nesting takes place in a hole excavated in the trunk of a tree. It is possible that the drumming the birds do to attract a mate also serves the purpose of identifying a suitably soft place to make the hole. The female lays 5–7 cream-coloured eggs, and both birds share the incubation and feeding duties.

MAGPIE

Pica pica

This distinctive black-and-white bird is a member of the crow family. Seen close-up it has a metallic green and purple sheen. It has an 'arrogant' gait. The sexes are similar but in the breeding season females can be told apart from males by their bent and damaged tail feathers, which occur because they alone incubate the eggs. Magpies live in close proximity to humans and consequently figure in quite a lot of folklore, children's stories and poems.

LENGTH 44cm.

HABITAT AND DISTRIBUTION Breeds around farms and villages, and increasingly common in urban areas. Widespread in much of Europe and Asia, with several subspecies within the range.

Magpies have a 'bullying' manner and will chase other birds away from food.

DIET This intelligent bird stores food for winter and breaks it into manageable sizes for its young. Magpies also wash food and are known from experiments in captivity to be able to count, so it is little wonder that they are so successful. They can take all kinds of food from bird tables, and although they are usually wary they get quite bold in some places, and may bully other birds away from feeders.

NESTING Magpies stay within a territory and pair for life. The nest is built high up in a fork between branches of a tree. It is a domed structure of twigs and leaves held together with some mud and lined with moss. There is usually one clutch a year. For its size the Magpie's eggs are small; they are pale blue with brown-grey spots, and take about 22 days to hatch.

This Magpie is on the lookout for food on the ground; the long tail is an obvious feature.

JAY

Garrulus glandarius

This colourful buff-pink bird is a member of the crow family. Despite its striking plumage it is not that easy to see as it is quite shy and wary; often just a flash of colour or the white rump is all that can be seen as it dashes between trees. It is usually the harsh, rasping shriek that gives away the bird's presence, although in some parks, gardens and picnic areas Jays overcome their shy ways and steal food.

Jays can raise their crown feathers as a threat or in display. They have a patch of blue-and-black feathers on the wings that were once prized by fly-fishermen. This led to them being hunted, with a consequent reduction in numbers; they now seem to be on the increase in urban areas more than in the wider countryside.

LENGTH 34cm.

HABITAT AND DISTRIBUTION Forests and parks throughout Europe and Asia and down into the Indian regions as well as north-west Africa.

DIET The diet includes seeds, nuts, acorns, invertebrates, small lizards and small mammals, as well as the nestlings of other birds. In a garden Jays are attracted to fat and suet cakes, peanuts and mealworms. They often store food for winter, burying caches of acorns and other nuts, including peanuts, in several places.

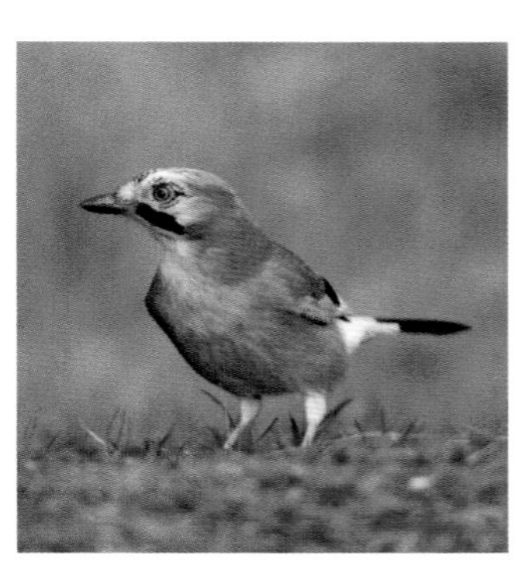

A Jay showing that distinctly blue-coloured wing patch.

NESTING The nest is a big structure of twigs lined with moss and hair, in which 4–6 eggs are laid and incubated by the female. The male attends to feeding her, and both adults feed the young.

Jays feed on acorns and they store them in winter. This one is searching an oak tree for them.

JACKDAW

Corvus monedula

The Jackdaw is one of the smaller members of the crow family. It is black overall with a silvery-grey head and nape, and grey eyes. It is a cocky bird with a strutting 'arrogant' manner, but is quite wary around people. Jackdaws are gregarious and vocal, and are usually seen in small groups searching for food on farmland and rocky cliffs, often venturing into gardens as well.

LENGTH 35–40cm.

HABITAT AND DISTRIBUTION Fields, woods, farmland and towns across Europe.

DIET The diet includes insects, seeds, vegetable matter and earthworms, as well as the eggs and chicks of ground-nesting birds. Being an opportunist, the Jackdaw is drawn to gardens, where it will eat most bird feeds; it is especially fond of fat and suet cake.

A Jackdaw in flight; other than the grey neck it is all black.

NESTING Nesting usually occurs in colonies, with nests being built in cavities in trees, buildings and rocky cliffs, as well as the insides of chimneys. Jackdaws are not sexually mature until their second year, but they pair for life, usually having just one clutch of young each year. The female incubates the 4–5 eggs, which are pale blue or blue-green speckled with olive-grey. In times when food is in short supply the last of the chicks to hatch will not be fed, ensuring that some of the young survive.

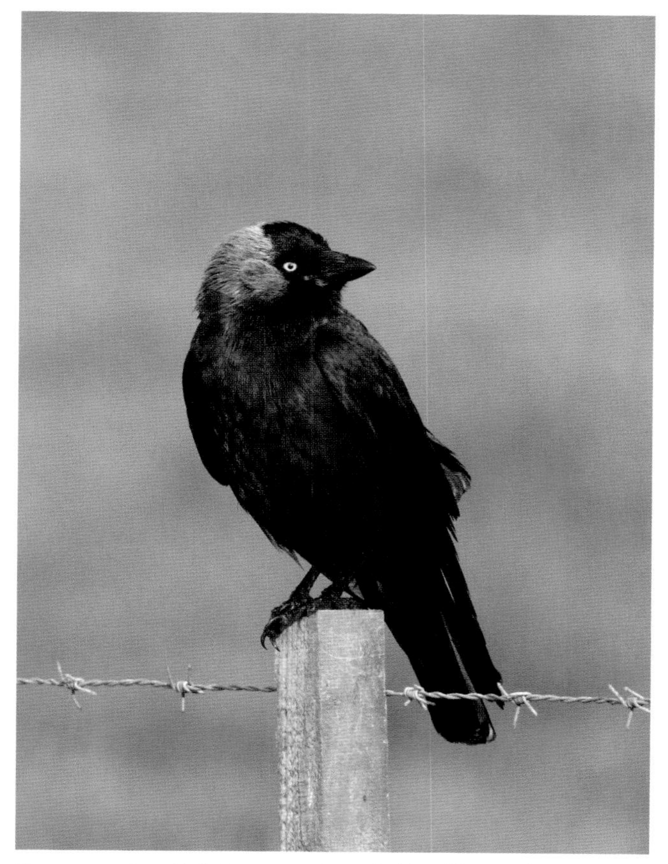

The grey neck and bright eye are the main identification features.

CARRION CROW

Corvus corone

This species is often seen over gardens and on rooftops, as well as at motorway service stations and leisure complexes. It is successful at exploiting any opportunity to scavenge food, and is never far from humans. Black all over with a purple-green 'oil-on-water' glossiness to its plumage and a shiny black bill, this crow is a smart bird with an 'arrogant' posture and attitude. It is also wary and quick to fly if threatened, which is why it is rarely seen dead by the roadside despite being a species that profits from road kills. Usually solitary but also mixing in loose groups with other crows, including Rooks *Corvus frugilegus*, where there is a source of food, Carrion Crows are noisy and intelligent. The Hooded Crow *Corvus cornix* replaces this species in eastern and northern Europe, and is common in parts of the UK.

LENGTH 47cm.

HABITAT AND DISTRIBUTION Coasts to mountains and towns across central and western Europe. In the UK, the Hooded Crow is common in north-west Scotland and Ireland, and rare on the east coast of England.

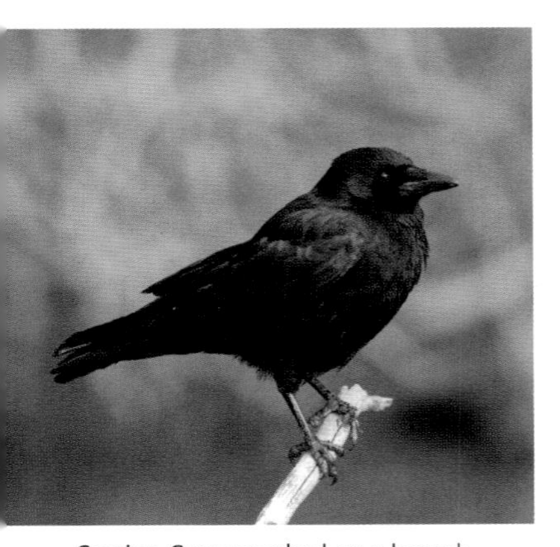

Carrion Crow perched on a branch.

DIET The diet includes earthworms, insects, nuts, seeds, grains and carrion, as well as the eggs and chicks of other birds.

NESTING Builds a large nest of twigs in a tall tree, or on cliffs, pylons or buildings, and usually raises 3–4 young.

The Carrion Crow is completely black, and can only be confused with a juvenile Rook.

GOLDCREST

Regulus regulus

This dull greyish-green bird, with its pale buff-grey belly and black-outlined yellow crown-stripe (orange-centred on males) is the smallest UK bird. Goldcrests do not stay still for long, always flitting around and often hovering as they busily search out their prey, at the same time uttering a quiet but high-pitched twittering noise and a *zi-zi-zi* call. They can be difficult to view well enough to separate them from the very similar Firecrest *Regulus ignicapilla*, which is brighter with a black eye-stripe and bold white supercilium, purely because they are so active.

LENGTH 9 cm.

HABITAT AND DISTRIBUTION Areas with pine trees are generally the best places to see Goldcrests, but they are widespread and fairly common in the UK and much of mainland Europe, and there are nine subspecies. They do migrate, and birds that are mainly from the north move for the winter months and swell populations in the south of their range. Adults and many juveniles are on passage in autumn, and are easiest to see at this time.

A Goldcrest actively searching for small insects.

DIET Feeds on tiny insects, spiders and grubs on the undersides of leaves. Goldcrests do visit garden bird tables, especially when the weather is harsh. They feed on fat cakes, especially those in hanging wire-mesh feeders, and eat mealworms, particularly small ones. They may also eat finely chopped sultanas, grated peanuts and cheese.

NESTING Nests in evergreen and mixed woodland, parks and gardens. The nest is a complex small structure made up of lichens and moss held together with spiders' webs and lined with hair, and is suspended on the end of a branch. The female alone incubates her clutch of 7–12 pale buff-white, brown-speckled eggs, and both parents attend to the young.

The yellow crown with black edges is distinctive; the rarer Firecrest is brighter above and has a bold white supercilium.

LONG-TAILED TIT

Aegithalos caudatus

Half of this pretty bird's length is in the tail. Long-tailed Tits are tiny, seemingly excited birds that show up in small flocks. Mainly black and white with pink and pinkish-grey on the wings and flanks, and short, stubby bills, birds of both sexes look similar. There are 19 subspecies that vary in markings; birds in northern continental Europe have white heads, and this race sometimes shows up in the UK.

LENGTH 13–15cm.

HABITAT AND DISTRIBUTION Woods with bushy undergrowth, hedges and gardens across most of Europe.

Note the orange eyelid; on juveniles the eyelids are dull red.

In winter Long-tailed Tits go around in small flocks of up to twenty birds.

DIET Feeds on insects and the eggs and larvae of butterflies, as well as buds and seeds. Suet cake, peanuts, sunflower kernels, fat balls and mealworms attract them to garden-feeder sites, and they can be regular visitors to water. They are especially good to watch as they vigorously splash in birdbaths and look bedraggled while they preen.

NESTING The nest is built in thorny bushes such as Gorse or Bramble. It is a globular construction made from small twigs, spiders' webs, lichen, moss and feathers. The female lays 5–16 eggs, and two clutches are usual per year. The eggs take 12–14 days to hatch and a similar time to the young fledging. Interestingly, other adult Long-tailed Tits, as well as the parents, help with feeding the young.

BLUE TIT

Cyanistes caeruleus

The lovely Blue Tit is unmistakable in its blue-and-yellow plumage. It has a bright blue crown, a blue tail and wings, and a black stripe down the yellow breast. These skilled little acrobats hang upside-down and cling on to the flimsiest of perches while they feed. Garden and birdfeeder regulars, they are usually one of the first species, along with their relatives and rival hole nesters Great Tits, to find and exploit new bird-feeding sites and nest boxes, often investigating them within days after they have been put up. Blue Tits are quick to learn and able to communicate what they have learned between each other – in the past they learned how to peck through metal milk-bottle tops to get at the cream on the milk inside.

LENGTH 12 cm.

HABITAT AND DISTRIBUTION Common in woodland, parks and gardens throughout Europe except Iceland and northern Norway. Nine subspecies are recognised, with a couple more in the Canary Islands that may be separated as full species.

A small, highly mobile and agile bird that will perch anywhere.

DIET Feeds on all sorts of bugs, spiders and caterpillars, as well as seeds and other vegetable matter. It frequently feeds in flocks of up to 30 in winter, often with other small bird species. The birds exploit a variety of human-prepared bird-food products, such as fat balls and fat cake, so are successful at surviving in close proximity to people. It is very important that you do not feed whole peanuts to garden birds (see p. 1) – Blue Tits are especially prone to choking either themselves or their young if they swallow them.

NESTING Blue Tits nest in April–May and can have large numbers of eggs, with 7–8 not being unusual, and there may be as many as 13. Two hens have sometimes been known to breed in the same nest box. The eggs are laid in a cosy cup of wool, moss, hair and feathers. They are white with reddish-brown speckles and take just two weeks to hatch. Three weeks later the young leave the nest.

The blue crown is very distinctive.

GREAT TIT

Parus major

This is the largest of the European tit family. It has a glossy black head with white cheeks, a black band down the centre of the yellow breast, a blue-green back and blue-grey wings with a white wing-bar. The male and female can be told apart by the width of the breast-band, which is wider on males. Great Tits appear to be quite smart and are usually one of the first birds to investigate new birdfeeders. They call all the time, making a great variety of calls: *teachur-teachur* is the most used, as well as a sharp, repeated *chink*.

Great Tits are defensive and aggressive to other birds, often squabbling and kicking out hard with their feet. They frequently hold food items against where they are perched with their feet, and peck pieces off with their sharp beaks. Because they are intelligent birds that can solve problems and use their acrobatic ability to gather food, they are an important study species in ornithological science because they help us to understand how birds adapt to man-made changes to the environment.

LENGTH 14cm.

HABITAT AND DISTRIBUTION Widespread in woodland and gardens across Europe except the far north.

DIET Feeds on seeds and fruits, and spiders and insect larvae in the breeding season. It is a regular garden visitor, eating sunflower seeds, peanuts and fat at birdfeeders.

Water is especially needed in winter.

NESTING A regular nest box user, making a moss-and-grass cup lined with hair inside a box and laying 7–15 eggs, sometimes twice in a season. The female does the incubating, only leaving the nest to drink and bathe, and the male feeds her. Both parents feed the young.

Great Tit showing the breast-band; the sex and race can be determined by the width of this.

CRESTED TIT

Lophophanes cristatus

This is a very easy bird to recognise because of its black-and-white crest. It is less colourful than other tits in that it is mainly grey-green above and white below, with buff flanks and belly. Like other tits it is quite 'chatty', constantly making a *zee zee* contact call to others, particularly in winter as it moves through the trees in loose flocks (often with other tit species).

LENGTH 12cm.

HABITAT AND DISTRIBUTION Crested Tits are not migratory, so are not found in gardens outside their range. They occur predominantly in mature coniferous forests of mainland Europe and northern Scotland. They are increasing in number in some areas of Scotland, where they are expanding their range into plantations.

The buff colouring and raised crest make the Crested Tit unmistakable.

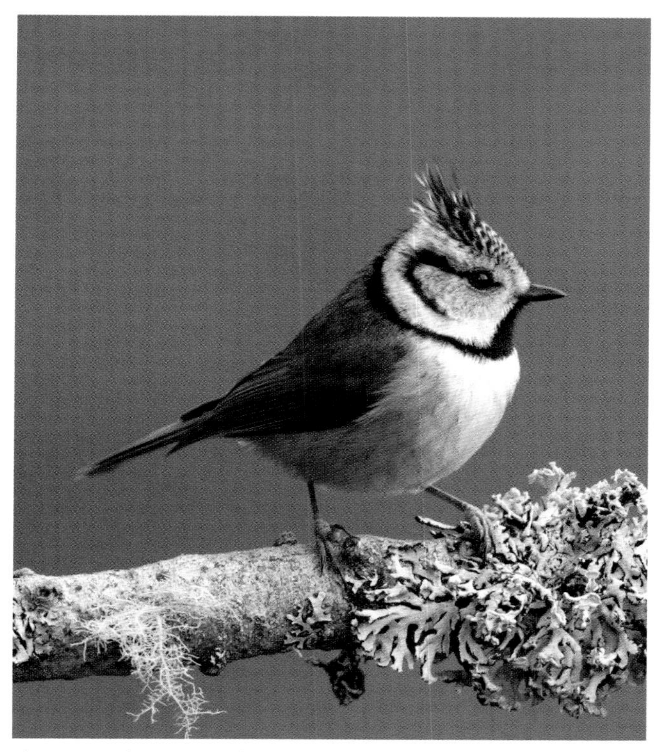

The UK population is confined to Scotland's pine forests.

DIET Feeds on a wide range of insects and spiders, as well as pine seeds, which are stored for consumption in bad weather. Crested Tits come to birdfeeders in places where they are found, feeding on suet cake and peanuts. Squirrel-proof feeders are a must as the birds compete with Red Squirrels in their range.

NESTING The nest is made in a hole, and the birds do use bird boxes. They make a cosy bed of lichen, moss and hair, and lay as many as eight glossy white eggs with purple-red speckles. The female incubates the eggs and both birds feed the young. The eggs and young do get predated by squirrels – a wire-mesh guard around a nest box helps to protect them.

COAL TIT

Periparus ater

A typical tit in shape and behaviour, the Coal Tit is not as brightly coloured as Blue or Great Tits. In common with them it has a glossy black head, but a bold white patch on the nape sets it apart and is a strong identification feature. The breast and belly are off-white with a buff wash, and there is darker buff colouration on the flanks. The sides of the face are off-white, as are the tips of the coverts, forming two noticeable wing-bars. The legs are lead-grey and the bill is black and finely pointed. In winter Coal Tits form up into loose flocks and mix with other tits and Nuthatches as they rove around woodland looking for food. They frequently make high-pitched *dee* contact calls.

LENGTH 11cm.

HABITAT AND DISTRIBUTION Widespread and common in woodland and gardens throughout Europe except the far north.

DIET Eats insects and seeds and is a frequent garden visitor, particularly favouring nuts.

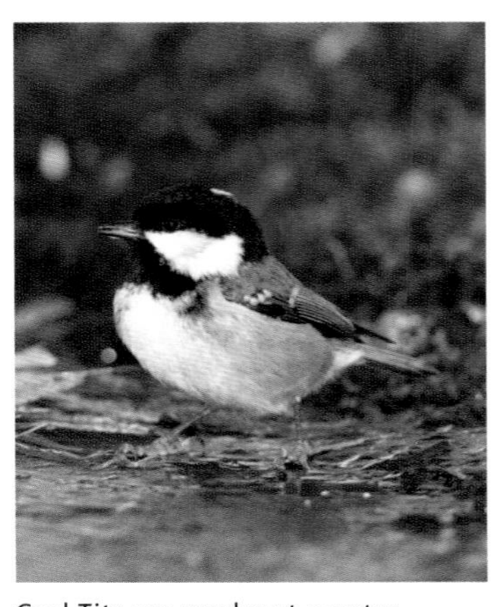

Coal Tits are regular at a water source, where they drink and bathe.

NESTING Nests in holes, and frequently uses nest boxes. It may also nest in gaps in stone walls and even odd sites such as large nests of other birds and squirrel dreys. Both birds line the nest with moss, hair and grass. The female lays and incubates 7–11 white eggs with reddish-brown speckles. Both birds feed the young, and they usually have just one brood per year.

The white patch on the back of the head of this species is shown well in this picture.

WILLOW TIT

Poecile montanus

This is a bit of a puzzle bird because it is so very like the closely related Marsh Tit. It is only the songs and calls that reliably tell them apart until one is familiar with the few diagnostic plumage differences. The Willow Tit's song is a *pee-oo-pee-oo* and the call is a wheezy, nasal *tchay-tchay-tchay* or a nasal *zee, zee, zee* (the Marsh Tit makes ringing *pitchoo* and *chicka-dee-dee-dee* sounds).

There are also visible differences between the species. The scruffier-looking Willow Tit has a pale patch on the wing and a bulging nape that gives the back of the neck a heavy look. It has a larger black bib and the black on the head is matt, lacking the glossiness of the Marsh Tit's head. The base line of the black cap is also different: on a Marsh Tit it runs in line with the centre of the eye, and on a Willow Tit it lines up with the bottom of the eye. As a general rule of thumb if you see a Willow Tit on your garden feeders, it will be a Marsh Tit – Willow Tits prefer damp woodland with birches and alders, are sedentary and rarely visit gardens.

LENGTH 12cm.

HABITAT AND DISTRIBUTION Forest, scrub and parks in Europe apart from the south-west. In the UK resident in England, Wales and s. Scotland, but recent catastrophic declines in many parts of the country make it even more unlikely to see one in your garden.

DIET Like other members of the tit family, Willow Tits mostly eat insects. In autumn and winter they eat berries and seeds, often storing a cache of food for bad weather.

NESTING Nests in cavities, with (usually) the females excavating the holes themselves in soft dead wood. The cavity is lined with hair, moss and feathers, before 6–9 (sometimes up to 13) glossy, white-and-brown speckled eggs are laid. Both adults feed the young, but only the female incubates them. Willow Tits are beginning to use nest boxes. They usually have to be filled with compacted wood chippings in which the birds can excavate a hole that suits them.

Identifying a Willow Tit in Britain is one of the toughest of challenges. The presence of a pale panel in the wing and a larger black bib are good pointers.

MARSH TIT
Poecile palustris

Marsh and Willow Tits are difficult to tell apart (see p. 92). Both species are sandy-brown above and buff below, and both have black heads and bibs. The key things to look for in Marsh Tits are a small, neat black bib, white cheeks and a glossy black cap; the wings also lack the pale wing-patch of Willow Tits. However, the Marsh Tit is most easily identified by its ringing *pitchoo* call, which sounds like a high-pitched sneeze, and a frequently uttered, more trilling *chicka-dee-dee-dee*.

LENGTH 12cm.

HABITAT AND DISTRIBUTION Mainly deciduous, older, established woodland and sometimes gardens in much of central and western Europe.

DIET Feeds mainly on insects, as well as seeds, berries and beechmast. It is a regular at feeders in areas adjoining the woodland in which it is found. It feeds on the ground more frequently than other tits. Seeds and fat cake can attract it to a garden, and it also favours live mealworms.

Male and female Marsh Tits can only be told apart by taking detailed measurements.

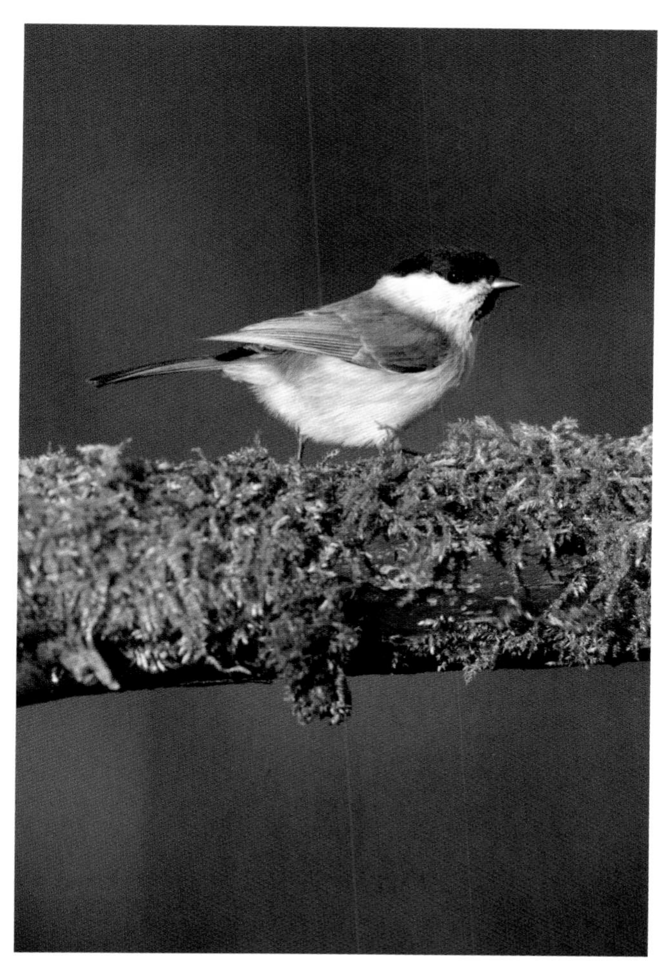

Marsh Tits come to feeders much more readily than Willow Tits.

NESTING Like other tits this is a hole nester and will use nest boxes regularly. The nest is lined with grass, moss, hair and feathers. The female lays 5–11 eggs, which are glossy white with brown marks. She incubates the eggs and the male joins her to feed the young.

SKYLARK

Alauda arvensis

A bird that appears to epitomise happiness, in spring and summer its joyful song is the theme tune of the countryside. Sadly it has declined due to the use of pesticides and certain farming practices, with the population having been reduced to about 10 per cent of what it was 30 years ago. A big conservation effort is under way to halt the decline.

This is a large lark, almost as big as a Starling. It is mainly brown with dark brown streaks above, and a paler buff-white with dark streaks below. The tail has prominent white outer feathers.

HABITAT AND DISTRIBUTION Meadows, heathland, salt marshes and moors; also occurs on arable farmland. Skylarks can be seen across Europe and Asia as well as areas of North Africa. They have been introduced to Hawaii, Australia and New Zealand, where – unlike in Europe – they are doing quite well.

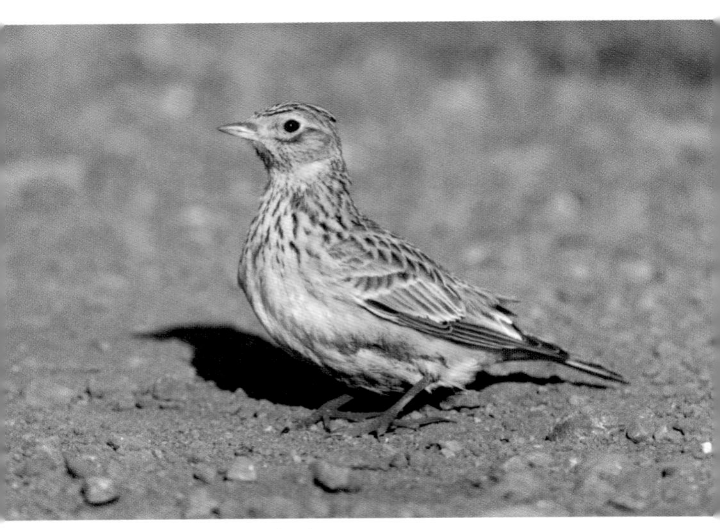

Skylarks regularly dust bathe.

DIET Feeds on seed, grains and insects. Also eats sunflower kernels and mealworms, live or dried. It is not a common garden visitor, but will come to food on the ground in gardens near its habitat at times of bad weather.

NESTING Nests on the ground, usually building a cup-shaped dry-grass structure lined with hair and moss. There are generally two clutches annually, each of 3–5 eggs, which are off-white with heavy brown and/or olive spots. The female incubates the eggs, and both the adults feed the young.

Skylarks raise their crest when alert or singing.

HOUSE MARTIN
Delichon urbicum

Like other hirundines, such as Barn Swallows and Sand Martins, House Martins are summer visitors to the region, arriving in Europe in late March to early May. Although the House Martin looks black, when seen properly the beautiful steel-blue colour is revealed. Its undersides and rump are white. The feet are tiny and black, and are partly hidden by white feathering on the legs. Although the bill looks small it has a wide gape; the bill parts are surrounded by stiff hairs that aid the birds in catching prey. Other than when the birds are collecting mud on the ground or perched on their mud nests, usually under the eaves of a house but sometimes on the brickwork under bridges, prolonged views of them enabling recognition of the fine details are unlikely. The birds are quite noisy, making contact chirps continuously. The male's song is a string of twittering chirps.

At the end of summer House Martins may be seen gathering in flocks, often on telephone wires, before they migrate to Africa for the winter.

LENGTH 14cm.

House Martins only land on the ground to collect mud for nesting.

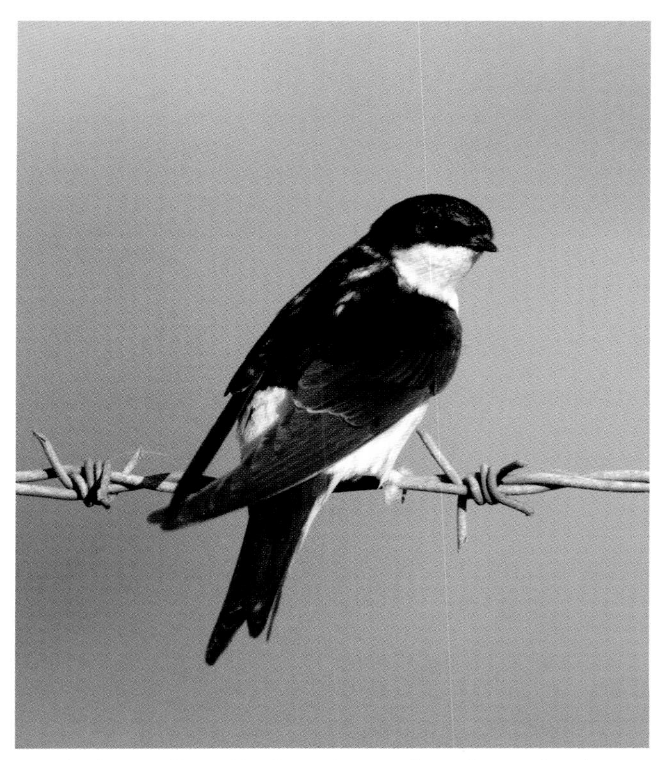

Later in the year House Martins gather on wires prior to migration.

HABITAT AND DISTRIBUTION Occurs in colonies in towns and villages, and on cliffs. Summer visitor and migrant to Europe, except the far north.

DIET Feeds on airborne insects.

NESTING Nesting is colonial with several nests being constructed close together. House Martins seem to pair for life. They usually have two broods a year. When the young leave the nest the adults carry on feeding them for a few weeks, generally in groups of several families. House Martins can benefit from being provided with nest boxes (see p. 12).

SAND MARTIN

Riparia riparia

The Sand Martin is one of the first hirundines to arrive ahead of summer. The species has white undersides, a brown back and a narrow brown band around the throat. It is very similar to the House Martin (which has a steel-blue back), but its flight is jerkier and it employs rapid turns as it chases airborne insects. Like House Martins, in late summer the birds gather in groups on telephone and fence wires, before migrating back to Africa.

LENGTH 12cm.

HABITAT AND DISTRIBUTION Open country with fresh water. Migrant across Europe except the far north.

DIET Feeds on insects such as flies caught on the wing. It is usually only seen in flight above gardens.

NESTING Sand Martins are colonial nesters, digging burrows in soft, sandy banks and cliff faces. The nest site is above or very close to water, and is often in quarries. The birds line the nest chamber with feathers, grass and hair. Once the 3–7 glossy white eggs are laid, both parents share incubation and feeding duties.

The bare legs and feet will help you to separate Sand Martins from House Martins.

Sand Martins rest on wire fences between hunting flights.

BARN SWALLOW

Hirundo rustica

Symbolic of summer, the Barn Swallow is a master of the air, catching its food on the wing with exquisite manoeuvrability. It is white below and a metallic royal blue above. There is a rusty-coloured patch on the forehead, chin and throat, and the long wings and tail streamers are very diagnostic of the species. It is a striking and easily identified bird. When Barn Swallows arrive in spring – usually late April to early May – they can be seen perched on wires, where they utter a tuneless chatter of twittering notes that sound a bit like the calls of Dunnocks but are more highly pitched.

LENGTH 19cm.

HABITAT AND DISTRIBUTION Breeds in farmyards and small-village gardens with surrounding open country. A summer visitor to northern Europe, it is found in most of the world. It migrates to the northern hemisphere in summer to breed, and returns to the warm south during our winter.

DIET Feeds on insects caught in flight low over fields and water.

Swallow collecting mud for nesting.

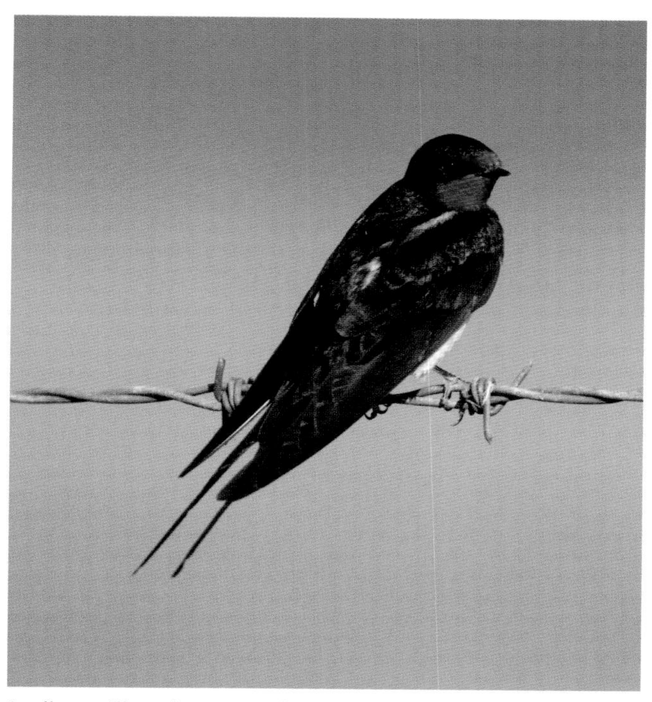

Swallows will perch to rest and continue twittering even with food or mud in their bills.

NESTING As soon as they pair the birds build a distinctive nest of mud, dung and saliva, bound together with short pieces of grass and animal hair. They often repair and reuse nests remaining from previous summers. Favoured sites for nests are the insides of barns, stables and outbuildings, as well as church porches and hides in nature reserves. The eggs are glossy white with ruddy-brown speckles. They hatch after about 15 days of incubation by the female. The male helps feed the young, and even after the juveniles have learned to fly the adults still feed them for a few more weeks; this is increasingly done in flight until they are able to fend for themselves. Swallows are great birds to view when they are collecting mud; a wet, muddy puddle can be provided for them to gather nesting material from.

COMMON CHIFFCHAFF

Phylloscopus collybita

This small warbler's name describes the song it delivers from a high perch during the summer months. Its best identification features are the brown-washed, olive-green upperparts (it can be confused with the Willow Warbler, which is generally a slightly brighter greenish-yellow), off-white underparts blending to yellow on the flanks, the off-white supercilium, and the thin, dark, pointed bill and dark legs. The song is the best identification feature, but it also uses a high-pitched *hweet* contact call that is frequently uttered and quite distinctive once you are familiar with it.

LENGTH 11cm.

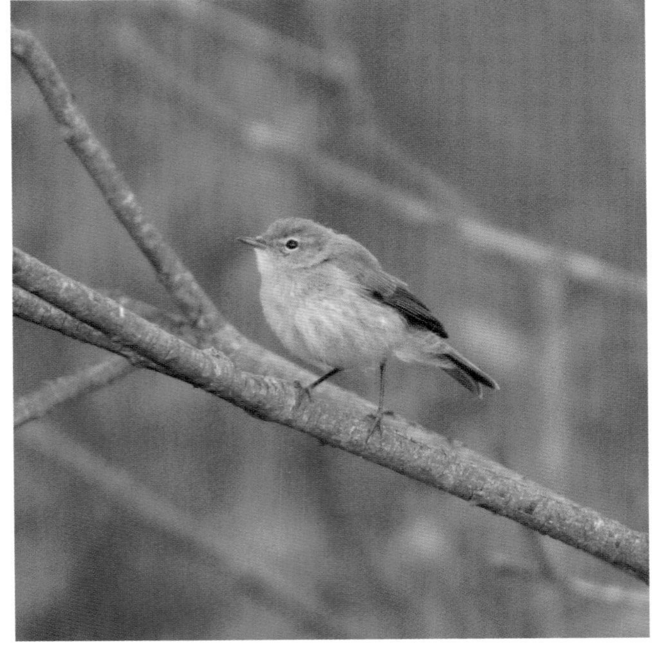

A juvenile bird: pale colours but it still has dark legs.

An adult Chiffchaff searching for insects and spiders to eat.

HABITAT AND DISTRIBUTION Open deciduous woodland with some scrub. Although this is mainly a summer migrant to central and northern Europe, increasing numbers stay in Britain during winter. Some of these birds are subspecies that are more usually found elsewhere in Europe and show slight variations in markings.

DIET The Common Chiffchaff is usually seen busily flitting around, flicking its tail, while it diligently searches out small insects to eat in the trees and bushes in a garden. It may also come to drink and bathe in a birdbath.

NESTING Once pair bonded the male defends his territory, even chasing off other females. He has little to do with nesting, and only has some involvement with feeding the young, mainly leaving the female to do much of the work. The nest is usually quite close to the ground in dense vegetation such as Nettles and Bramble bushes.

WILLOW WARBLER

Phylloscopus trochilus

A typical leaf warbler in shape and size, this species is greenish-brown above and a yellowy off white below. It has no wing-bars, and pale pinky-yellow legs and bill; the leg and bill colours are the main visual diagnostic points for separating it from the very similar Common Chiffchaff. A close-up look reveals that the Willow Warbler lacks the latter bird's eye-ring. However, the song is also different in that it is a melodic whistle of rising notes that slowly fall away, repeated several times, and usually sung again after moving to a nearby open perch. The contact call is a slightly longer *hoo-eet*, very similar to, but distinguishable from, the Common Chiffchaff's wheezier *wheet* contact call.

LENGTH 11–12.5cm.

HABITAT AND DISTRIBUTION Upland birchwoods and other deciduous woods, parks and large gardens. Willow Warblers arrive to breed in Europe and Asia from Africa during April. There are three recognised subspecies across the range – birds get slightly browner and greyer towards the eastern areas.

Willow Warblers will often perch out in the open and snap at passing bugs. Note the pale legs.

DIET Feeds on insects such as aphids and caterpillars, and can often be seen fluttering and hovering as it grabs food from the undersides of leaves. Although the birds may take mealworms, providing water is a better means of attracting them. They drink and bathe several times a day, so will regularly come to a birdbath or pool, especially in a long, dry spell when there are few places to drink.

NESTING The nest is a domed structure of grass and moss lined with hair and feathers, usually erected close to the ground in the cover of tangled vegetation. The eggs are glossy white with fine ruddy speckles. Willow Warblers may have two broods annually of 3–8 eggs each, which the female broods herself. The male feeds her and both adults feed the young when they hatch.

Willow Warblers diligently search the leaves and buds for small insects.

BLACKCAP

Sylvia atricapilla

A very smart warbler, the Blackcap has olive-grey upperparts and pale grey undersides. The male has a crisply defined black crown, while the female's crown is brick-red. Blackcaps are territorial, and their territories are established and maintained by the male. He raises his crown feathers and slowly flaps his wings while singing loudly to attract a mate.

LENGTH 7–8cm.

HABITAT AND DISTRIBUTION Open woodland, shrubby areas with trees and gardens. The Blackcap is less migratory than many warblers or migrates in an east–west direction. Many birds stay during the cold weather, and records suggest that German-bred birds arrive in the UK to live alongside local birds in winter.

Male Blackcaps are very distinctive, with contrasting black crowns and greyish plumage.

Females are similar to the males, but somewhat browner with reddish-brown crowns.

DIET The diet is mainly insects, nectar and pollen during summer, and an increasing amount of fruits in winter. Interestingly, Blackcaps squeeze out the seeds of berries before eating the soft pulp and can be important in the propagation of species like Mistletoe. They also eat Elderberries, so planting these in a garden provides them with food as well as affording good views of the attractive birds. Despite the frequency of Blackcap visits to bird tables little is known about what foods they prefer, but they do come to fat cake containing berries and dried insects, and feed on windfall apples.

NESTING The male may build two or more nests for the female to choose from, usually in deep cover in Brambles and quite low near the ground. The female lays 4–6 buff-coloured eggs with brown-grey blotches and spots.

COMMON WHITETHROAT

Sylvia communis

This bird's scratchy song is a common sound of the summer countryside. It sings from an open perch on top of a hedge, but at all other times it skulks in the undergrowth and can be difficult to see well. Both sexes have warm chestnut-brown wings, a grey-brown mantle, a pinky-buff breast, and a white throat, belly and outer-tail feathers. The male is noticeably brighter overall and has grey on the crown, ear-coverts and nape; the female is brown on these features.

LENGTH 14cm.

HABITAT AND DISTRIBUTION Open woodland, hedgerows, scrub and gardens. A summer visitor to the European and Asian areas where it breeds, and a winter visitor to central Africa and Pakistan, the Common Whitethroat usually arrives in the UK in late March or early April.

DIET An insect eater for much of the year, the species also eats berries, especially when building up strength before migration. In gardens it may take mealworms, and because it likes to bathe and drink a few times a day, it can be regular at a birdbath.

Male Common Whitethroats sing from tops of bushes in spring.

NESTING The deep, cup-shaped nest is built by the male using grassy roots. It is sited in cover close to the ground, and he may build more than one nest. The female chooses the neatest one and lines it with feathers and hair. She lays 3–7 pale blue-green eggs speckled with olive-grey, and she and the male take turns to incubate them. Both parents feed the young in the nest and for several days after they leave.

Whitethroats bathe a lot, sometimes four or five times a day.

LESSER WHITETHROAT

Sylvia curruca

Slightly smaller than the Common Whitethroat, the Lesser Whitethroat is overall greyer in colour and has milky-grey underparts, whereas the Common Whitethroat is a richer brown-grey on the mantle and has a pinkish shade to the underparts. The Lesser has a shorter tail and white outer-tail coverts, and does not have as obvious an eye-ring. It is usually a skulking bird that stays in deep cover, making its presence known with a sharp *chit* or *chack* call. The song is a rattle of notes that is short but far-carrying.

LENGTH 13 cm.

HABITAT AND DISTRIBUTION Farmland with trees and hedges, parks, woodland edges, large gardens and scrub. A summer visitor to Britain across Europe into Asia, it winters in Africa.

DIET Feeds mainly on insects, including flies and caterpillars in summer. In autumn and winter it also feeds on berries and other fruits.

NESTING The nest is constructed from twigs, grass and hair, and is built low down in scrubby cover. The female lays 4–6 eggs twice a year, and both adults share the chores.

A very active warbler, constantly searching flowers and leaves for food.

The lesser whitethroat is much greyer than its commoner relative.

WAXWING

Bombycilla garrulus

In recent years Waxwings have become a common winter spectacle as they flock to places such as supermarket car parks to strip Rowan and Cotoneaster trees of their red berries. It is probable that more of the birds come to the UK in winter because of the rise in out-of-town shopping and the types of plant used to ornament the environs around the stores.

About the size of a Starling, this smart bird is pinkish-beige in colour with a large crest that either lies flat and protrudes from the back of the head, or is held erect above the rust-coloured forehead. A black mask runs back through the eye and the bird has a small black bib. The rump is grey, running to a black tail that is tipped with yellow. The species gets its name from the wax-like 'fingers' on the secondary feather tips.

LENGTH 20cm.

HABITAT AND DISTRIBUTION In Europe breeds in coniferous forests in the far northern taiga. In winter flocks to parks and gardens with berries south to Britain and the Balkans.

DIET Waxwings are acrobatic feeders; they show up in gardens where there are suitable berry-bearing bushes or windfall apples. They also come to birdbaths to drink and bathe. Once the young leave the nest, nomadic flocks of sometimes several thousand birds form up to spend the winter months looking for food.

Look in bushes with winter berries for Waxwings.

NESTING A cup nest is made from twigs. There is just one clutch annually of 4–6 blue eggs with black speckles, which is usually incubated by the female.

The Waxwing is a very smart bird, similar in size to the Starling.

NUTHATCH

Sitta europaea

With a pale blue-grey back and a peachy-orange breast and belly, pale cheeks and black 'bandit's mask' eye-stripe, this is a great-looking bird, which looks 'pointed' at both ends (the northern European subspecies is white below, rather than orange). It is a joy to watch the antics of a Nuthatch. It runs up and (head-first) down tree trunks and stone walls, searching out prey to feed on.

LENGTH 14cm.

HABITAT AND DISTRIBUTION Mixed deciduous woods, parks and gardens with mature oaks. Nuthatches appear to be doing well, with a slight increase in population and a northwards spread in their range across Europe and Asia. In the UK they are absent from Ireland and Scotland.

Nuthatches are the only birds that walk down trees head-first.

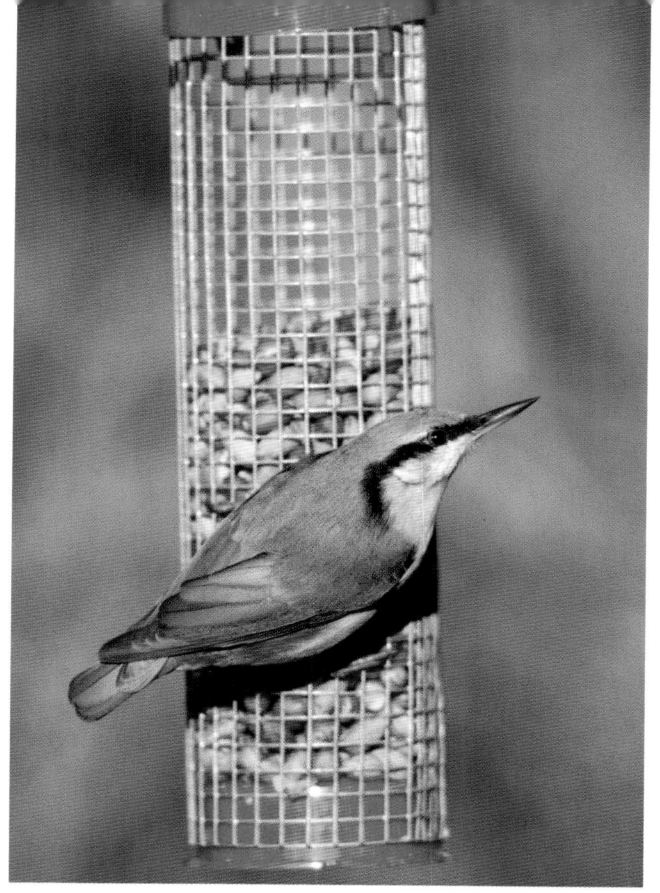

Nuthatches are often regular visitor to feeding stations.

DIET Feeds on invertebrates such as insects and larvae. Also eats a range of seeds, nuts and fruits. It takes mealworms, suet cake and peanuts at birdfeeders.

NESTING Naturally nests in holes and crevices in trees and rocky walls, often making the entrance hole smaller by plastering mud around it. Commonly uses nest boxes. The female lines the nest with dried leaves and bark chips. The eggs are glossy red spotted with white, and the female alone incubates them. The male joins her in feeding the young.

TREECREEPER

Certhia familiaris

This inconspicuous little bird has stiff tail feathers and looks like a tiny woodpecker as it creeps up tree trunks. It is brown with pale spots and streaks above, and buff-white below, with a rufous rump and pale brown legs and feet. The bill is long and curves downwards; it uses this to prize insects and larvae from gaps in bark. There are a few confusable species; Short-toed Treecreeper *Certhia brachydactyla* is almost identical, but has less white on the underside and separable only by its different song; it is rare in the UK.

LENGTH 13cm.

HABITAT AND DISTRIBUTION Woodland, parks, orchards and gardens with old trees with loose bark for nest sites. Found across Europe except the south.

DIET Probes bark crevices for insects and spiders. It is mainly insectivorous and not that common on birdfeeders unless the weather is harsh. Smearing suet, fat or peanut butter into a hole in a tree can attract it, and it may be a regular visitor to a dish of live mealworms, usually making a smash-and-grab visit and flying off with its prize.

Because of its cryptic markings it can be easy to overlook a Treecreeper.

NESTING Usually builds a nest behind the bark in a tree crevice, but sometimes uses a crevice in a wall. It occasionally uses nest boxes. The nest has a base of small twigs, grass, bark flakes and pine needles, and is lined with feathers, wool and hair held together with spiders' webs. The female lays 3–9 matt white eggs with small ruddy spots and incubates them. The male joins her in feeding the young.

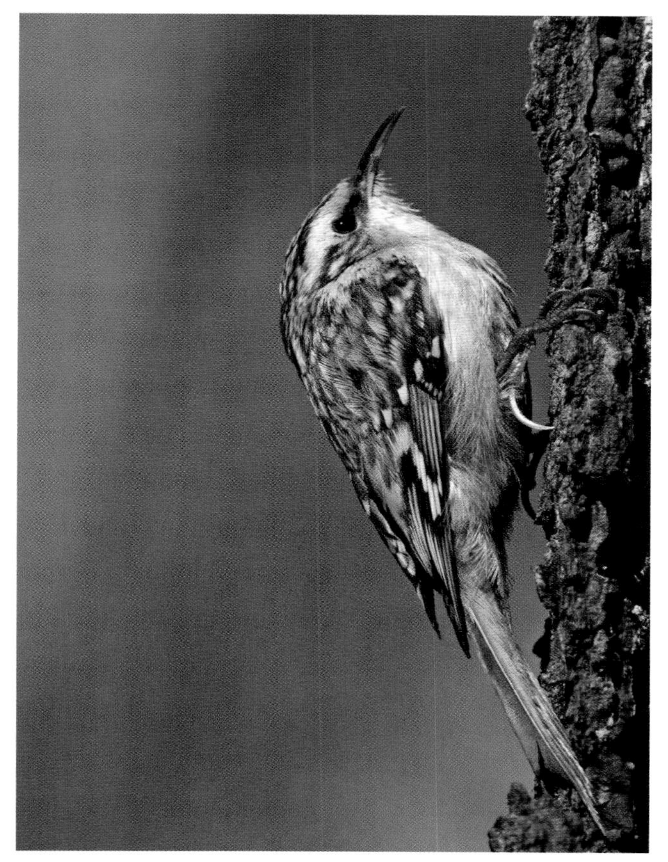

At close range, the upperparts are intricately-patterned. Treecreepers from northern and eastern Europe are whiter below than British birds.

WREN

Troglodytes troglodytes

A tiny bird with a very big voice, the Wren is known to many. It used to be featured on the backs of farthing coins, and still appears on Christmas cards. It is an endearing bird, scuttling about with its tail cocked high while searching for food, usually investigating tangled piles of branches, wood and scrub. The species is unmistakable – it is all-over brown in colour with dark barring, a needle-sharp bill and a pale supercilium, plus huge feet for its body size. The sexes are similar, and in flight the wings blur with rapid beats as the bird darts across between the cover of vegetation. The male is territorial and sings to declare territory ownership as well as to attract a mate, belting out a loud, rapid trill of notes as well as a sharp, scolding *teck teck* alarm call.

LENGTH 9–10cm.

HABITAT AND DISTRIBUTION Woodland with dense undergrowth, scrub, parks, gardens, heathland and moorland in most of Europe. Wren populations fluctuate depending on the severity of winter weather, but they do seem to recover well after falls.

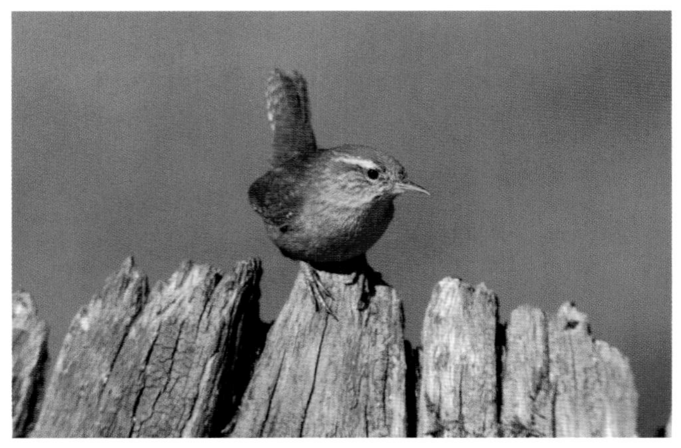

Wrens usually have their tails upright.

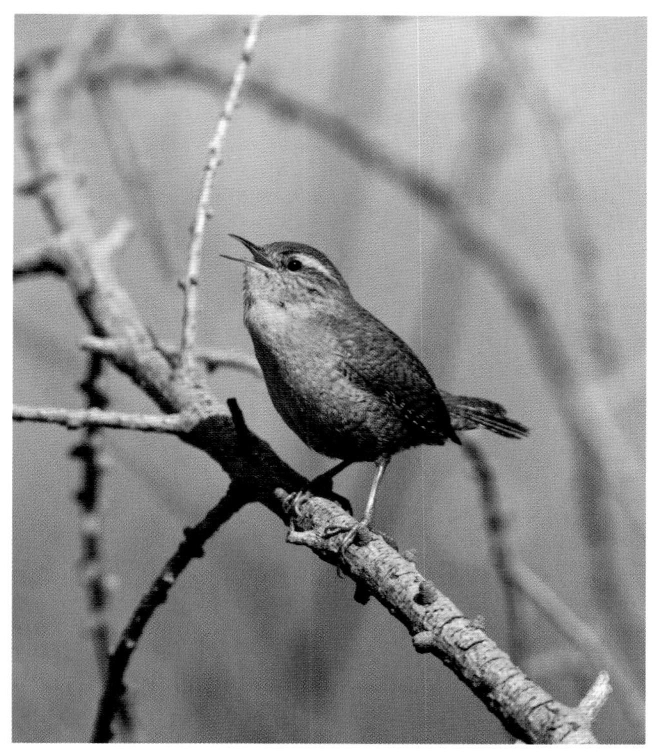

Wrens are territorial; they find a prominent perch to sing from so the song carries a good distance.

DIET Searches mouse-like for insects and spiders on or near the ground. May feed at bird tables, eating live insect food and fat and suet cake; also favours peanut butter.

NESTING Wrens nest in many places, using all kinds of nest boxes, sheds, crevices, trees and piles of junk. The nest is a round ball of twigs, grass and moss, lined with hair and feathers; it has an entrance hole in the side. The female lays 5–8 glossy white eggs with red-brown spots that look quite big for the size of the bird. The female incubates the eggs and both parents feed the young.

STARLING

Sturnus vulgaris

Starlings used to be seen in huge roosting flocks in Britain, but there are nowhere near as many now. They are still, however, quite common birds in town gardens. They must watch for food to be put out, because they arrive en masse to noisily scoff the lot very soon after, strutting around like avian bullies. They are cocky, arrogant and gregarious, and a joy to watch because they appear clever and intelligent.

Despite perhaps initially appearing like scruffy black birds, in fact Starlings are stunningly beautiful, with purple-and-green iridescence to their feathers, buff edges on the wing feathers, white speckles above and below, a sharp yellow bill, and reddish-brown legs and feet. Males and females are easy to tell apart, although at first glance they look very much alike. The base of the male's bill is pale blue, while the female's is pale pink; the male's iris is dark, barely separable from the pupil, and the female's is an obvious mid-brown colour. Starlings are vocal, uttering several sounds and having a song of their own, but they also mimic other birds, as well as telephone ringtones.

LENGTH 21cm.

In winter Starlings form large flocks before going to roost.

HABITAT AND DISTRIBUTION Widespread across Europe in all habitats, particularly near humans. Has recently decreased alarmingly in Britain, where it is now considered endangered.

DIET Starlings eat pretty much anything – seeds, fat balls, suet and kitchen scraps like uncooked pastry – so they soon get used to a feeding site, becoming quite tame. Away from gardens they often feed with flocks of Lapwings and Golden Plovers *Pluvialis apricaria*, picking up the food items that the other birds disturb.

NESTING The male builds the nest and the female lines it with hair, wool and feathers. Both birds incubate the 4–9 pale blue eggs. Female Starlings may also lay eggs in another Starling's nest, a strategy that ensures the survival of as many offspring as possible.

Adult Starlings have colourful plumage when the light is at the right angle.

BLACKBIRD

Turdus merula

A member of the thrush family, the Blackbird is one of the most common species in the UK, with the population increasing in winter when many continental birds arrive from further north. The male is a smart glossy black all over with a bright yellow bill and an inquisitive eye surrounded by yellow skin. It is often seen as it jerkily searches the lawns in parks and gardens for earthworms. The female is a duller bird, a warm-toned dark brown with a dark yellow-brown bill, and a lighter brown breast. Young females have some speckles like those of a Song or Mistle Thrush. The male has a rich, melodious song, and both sexes use a variety of sharp, scolding cluck sounds and alarm calls, the latter being voiced in particular when they have fledglings fresh from the nest.

LENGTH 25–29cm.

HABITAT AND DISTRIBUTION Woodland, gardens, parks and orchards across Europe.

DIET Eats a wide range of foods, including insects, worms and fruits. Frequently visits bird tables and is particularly partial to mealworms. Blackbirds are a gardener's friend in that they feed on 'pest' animals such as slugs, caterpillars and beetles, often approaching very close to take tasty treats from a garden as it is dug.

Male Blackbirds are distinctly different from females.

NESTING Breeding can take place in any part of a garden, in hedges, climbing plants, sheds or garages. The nest is a mud-lined cup woven from dried grass. The female lays 3–4 pale blue eggs that take 12–14 days to hatch. The youngsters leave the nest after another 18–20 days.

Overall the female is brown; some are quite pale and mottled on the breast like this one.

REDWING

Turdus iliacus

With a wingspan of 33–34cm, the Redwing is similar to the Song Thrush. It has a dark brown back, a buff-white breast with dark speckles, an orange-red patch on the flanks, a creamy-white supercilium above the eye and a pale moustachial stripe.

LENGTH 20–23cm.

HABITAT AND DISTRIBUTION Redwings are found right across Europe, west to Greenland and east right across Russia. They breed in the north of their range in birch and conifer woodlands. Several have been recorded nesting in northern Scotland, which is about as far south as they are in summer. In the winter, however, Redwings flock together and move south as the weather bites.

Redwings flock with Fieldfares and strip berries from hedgerows in winter.

DIET Redwings flock together with Fieldfares in winter as they feast on berry-laden hedgerows, or search short-grass fields for insects and worms. School sports fields and freshly tilled farmland are good places to see them. They may come to gardens containing Rowan or any other berry-producing trees. They are usually wary but lose their shyness in cold, frosty weather, when they come to garden feeders – mealworms and windfall apples and pears attract them.

NESTING Generally lays 4–6 eggs in a typical grass-cup nest situated near the ground.

Snowy weather will bring these birds to gardens in search of food and water.

FIELDFARE

Turdus pilaris

A lovely member of the thrush family, the Fieldfare is a medium-sized bird that is usually quite wary until the cold weather bites and its need to feed overcomes its desire to flee. The sexes are similar, with the birds sporting a rich brown back, grey on the back of the head and rump, and a dark grey-black tail. The breast is white with a reddish-brown wash towards the throat; this is speckled with dark, arrowhead-shaped spots. A handsome bird when puffed up to keep warm, the Fieldfare looks bold and bossy; it acts that way too, vigorously defending a food supply and chasing off rivals.

LENGTH 22–27cm.

Fieldfares will come to larger gardens and orchards in winter.

Some Fieldfares aggressively protect their food when the weather is harsh.

HABITAT AND DISTRIBUTION Breeds in northern Europe right across into Asia. Fieldfares are strong migrants that move south ahead of the cold weather in the winter, often in large flocks.

DIET As soon as they make landfall after their long migratory flights across water, Fieldfares fill the hedges and bushes, noisily feasting on the new crop of autumn berries while uttering sharp *chack chack* calls as they fly and feed. They are great birds to see at garden feeders in winter. They love windfall fruits such as apples, pears and plums, and also eat earthworms, mealworms and wax worms, as invertebrates and insects are naturally part of their diet.

NESTING The nest is a grassy cup in the fork of a tree.

SONG THRUSH

Turdus philomelos

Similar to but smaller than the Mistle Thrush, the Song Thrush is about the same size as a Blackbird but has a slimmer appearance. Both sexes are warm brown above and have a buff-white breast and belly with yellowish-buff flanks. The breast and flanks are covered in dark spots, which are fewer and smaller than those on the Mistle Thrush. Song Thrushes are often seen on lawns and in flower beds as they search out worms and snails, frequently standing erect on their pink legs with the head cocked to one side as they look for prey; however, they prefer to hunt in cover under bushes and hedges. Although they have declined in numbers, in the last few years they have been showing slight signs of recovery.

LENGTH 23cm.

HABITAT AND DISTRIBUTION Woodland, parks and gardens in much of Europe.

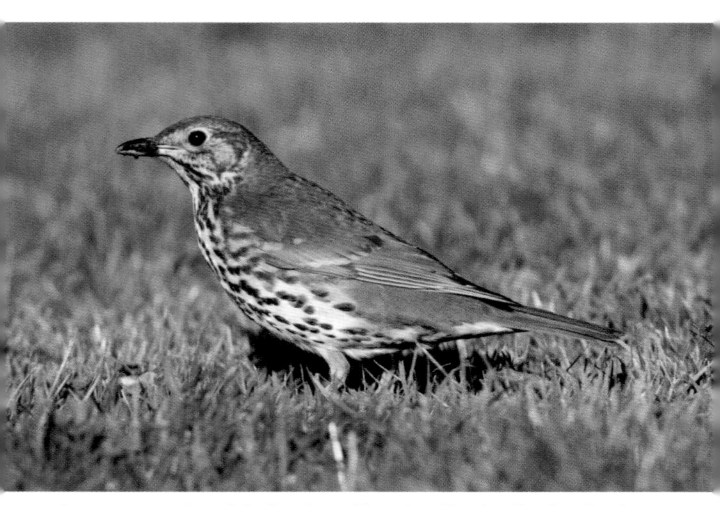

A common spring sight is a Song Thrush collecting food to feed its young.

DIET Feeds on worms, insects, berries and snails. A snail is taken to a favourite stone or wall and smashed against it to break open the shell. Recent research suggests that snails are not a main part of the bird's diet, but more of a back-up food when the ground is too hard to dig out worms. Song Thrushes come to garden feeders and eat suet cake, seeds and fruits such as windfall apples and pears; they also favour live mealworms. The birds can become quite tame – a well-placed stone and a supply of snails nearby will give great views of 'snail smashing'.

NESTING Nests in hedges and trees, and also in man-made sites such as greenhouses and sheds. The grass-and-twigs nest is built on a support (a tree bough or beam in a building), then lined with a mix of mud and the bird's saliva to set it hard. The 3–9 eggs are bright blue and glossy with a few black spots, and there may be as many as three clutches in a season.

In winter when insect food is scarce Song Thrushes come to bird tables or food left on the ground.

MISTLE THRUSH

Turdus viscivorus

Also known as the Stormcock because of its beautiful song, which was once assumed to precede bad weather, the Mistle Thrush is a larger and paler version of the Song Thrush and is the biggest of the European thrush family. It is easily separable from the Song Thrush by its grey-brown back and dirty pale-grey breast with large dark spots – the Song Thrush is a pale yellowish-buff on the breast and flanks – and its deep, undulating flight and chacking alarm call are very diagnostic.

LENGTH 27cm.

HABITAT AND DISTRIBUTION Woodland, parks and gardens across Europe. Common but declining.

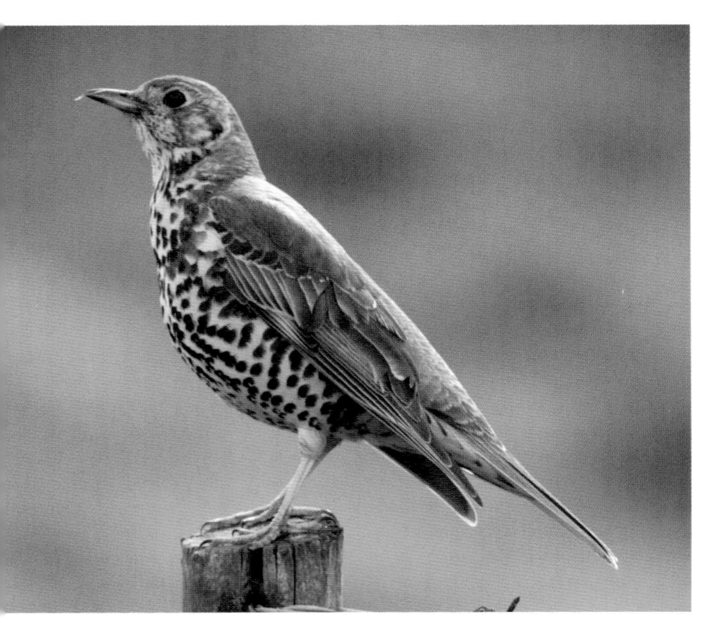

The Mistle Thrush is larger and greyer than the Song Thrush.

DIET Eats worms, slugs, insects, and also berries from many trees, including Yew, Holly, Rowan and hawthorn. Mistle Thrushes come to gardens mainly to collect natural food items, often searching lawns for worms. They love mealworms and in hard weather also eat fat cake.

NESTING The nest is a bulky cup in shape, constructed from small twigs, grass and mud, and lined with moss and soft materials such as feathers and hair. There are usually two clutches annually, each of 3–6 pale blue eggs with ruddy-brown spots. They are hatched by the female and the young birds are fed by both parents.

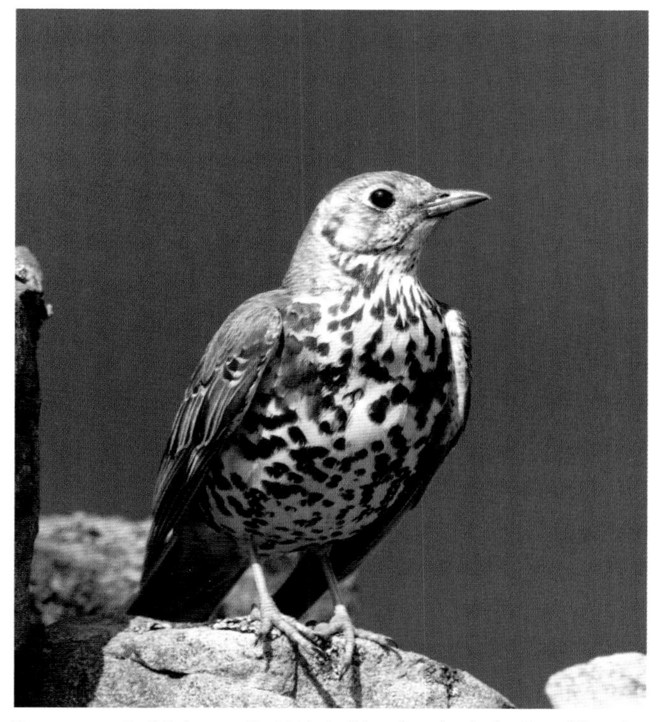

You can see in this image that it is bulkier than both the Song Thrush and Blackbird.

SPOTTED FLYCATCHER

Muscicapa striata

This bird is fairly inconspicuous and often only seen as it darts out to catch prey in mid-air – if one is close enough it is possible to hear the bird's beak snap shut. Its plumage is streaky brown on the back and crown, and pale on the breast and belly, with buff-brown streaks on the flanks and breast. The little feet and beak are black. These are easy birds to view if they are present in a garden. If a suitable perch is placed near to where flies gather the birds will use it to hunt from.

LENGTH 14cm.

HABITAT AND DISTRIBUTION Late-arriving migrants to Britain, usually showing up in mid-May and departing again to tropical Africa during August. They can be seen in gardens, parks and especially churchyards – anywhere that is quiet and still.

DIET Feeds on insects. Hunts from higher up in warm weather than in cold weather because the insects, butterflies and hoverflies it prefers to eat are lifted by the warm air.

Seen here on a churchyard headstone, Spotted Flycatchers are equally at home in well-established gardens.

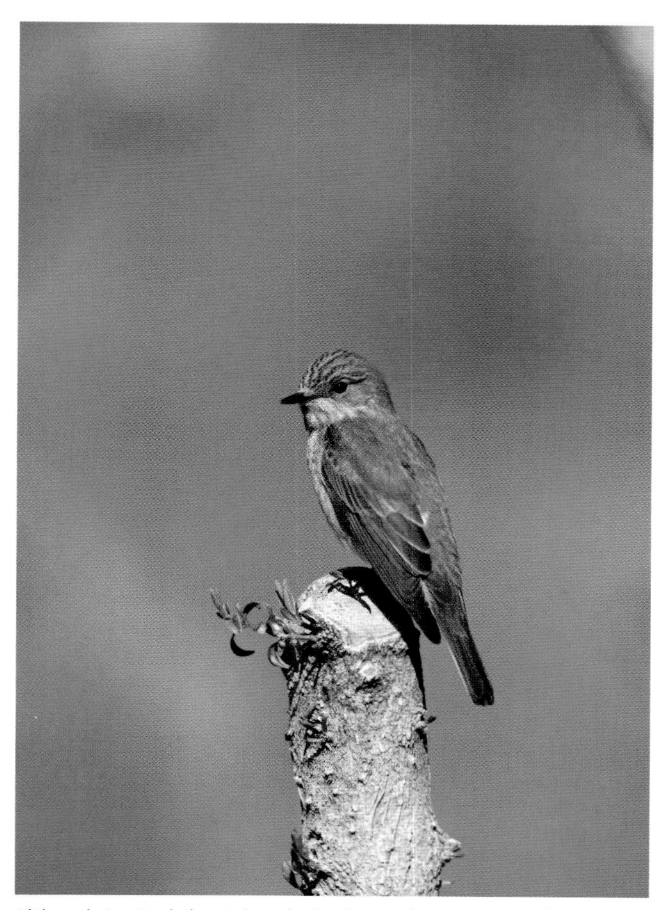

Although Spotted Flycatchers lack bright colours, they are fun to watch while they hunt insects.

NESTING The nest is built in a variety of sites, including cracks and crevices in trees and buildings, and nest boxes. The birds prefer the open-fronted type of nest box, especially if it is placed behind the cover of wall-climbing plants. Two clutches annually of creamy-white eggs with reddish blotches are laid. Only the female incubates them, and the male helps to feed the young.

PIED FLYCATCHER

Ficedula hypoleuca

The male has a black back and head, and white underparts. The female is a similarly marked buff-brown colour where the male is black. Both sexes have jet-black legs, feet and beaks. As they grow juveniles look similar to females, but have streaky markings on the back.

LENGTH 13cm.

HABITAT AND DISTRIBUTION This summer visitor to Britain is usually found in hilly woodland, especially oak, during the breeding season, but it can turn up anywhere while migrating. It begins to migrate south to West Africa in late August.

DIET The diet consists mainly of insects, which are collected from leaves and other vegetation and taken on the wing, making the birds difficult to see as they dart about in the trees.

NESTING This is a hole nester, usually employing natural crevices in trees and rocks, but increasingly using nest boxes. The female incubates the pale blue eggs and both adults feed the young.

Female Pied Flycatchers are less boldly marked...

...but the males are distinct; both are adept at catching insects on the wing.

ROBIN

Erithacus rubecula

A brown back and a red breast added to a tame nature is all the description this bird needs. It is the iconic bird of Christmas cards and calendars, and familiar to most people in the UK. The male Robin has a beautiful fluting warble song that some people mistake for the song of a Nightingale because it can be heard all night during spring. In winter the males and females both have a softer, more plaintive territorial song. They also make a number of other calls, including a sharp, scolding *tick* alarm call.

LENGTH 13 cm.

HABITAT AND DISTRIBUTION Woodland, gardens, parks and forest edges. The range stretches across Europe to western Siberia and down into North Africa. Plumage variations in the area suggest that there are six subspecies with two more very distinct ones in the Canary Islands.

DIET Feeds on insects, worms, caterpillars and grubs. In gardens Robins eat anything that is supplied, with cake, dried fruits and uncooked pastry being favourites. They also eat sunflower kernels, seeds and mealworms, and with a little patience soon become hand tame.

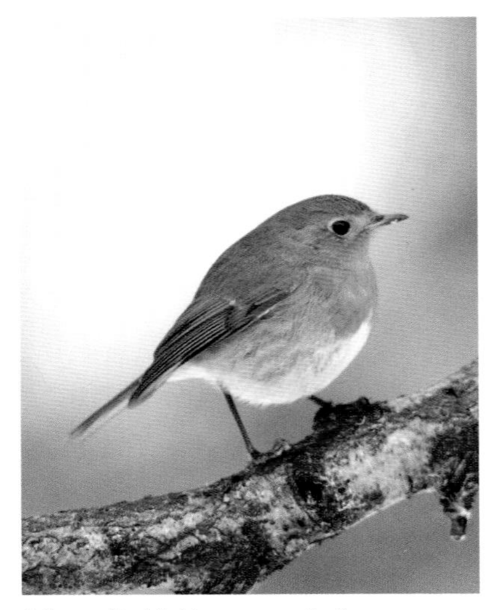

A favourite bird to many, whether a birdwatcher or not!

NESTING Robins nest in open-fronted nest boxes, although they naturally nest in holes and crevices near the ground. They are quite adventurous with nest sites in the vicinity of humans, having nested in cars, tractors, buildings, plant pots in greenhouses and between paint tins in sheds. The nest is a grassy cup lined with moss, hair and wool, and there may be up to three clutches of 3–7 eggs each year. The female incubates the eggs, and the male joins with the feeding of the nestlings and mottled brown and golden-brown juveniles.

Robins always sound cheerful when singing.

NIGHTINGALE

Luscinia megarhynchos

Renowned for its fabulous song, the Nightingale is a plain brown passerine that looks like a Robin, but without the red breast, and is in the same chat family. It is plain brown above with a rich red-brown tail, and paler below with a big, black, beady eye surrounded by a pale eye-ring. Males set up territory after their migration to Britain, then sing, often from deep cover where they can be difficult to see. The song is a rich and varied mix of sweet and melodious notes interspersed with an almost frog-like, croaking *churr* sound. Paired males stop singing while unpaired ones carry on until about late May, often throughout the night.

LENGTH 17cm.

HABITAT AND DISTRIBUTION Lowland deciduous woodland, particularly that containing coppice or low, bushy undergrowth. Summer visitor to southern England and across Europe except the north. Males arrive slightly ahead of females in mid-April from their wintering grounds in central Africa.

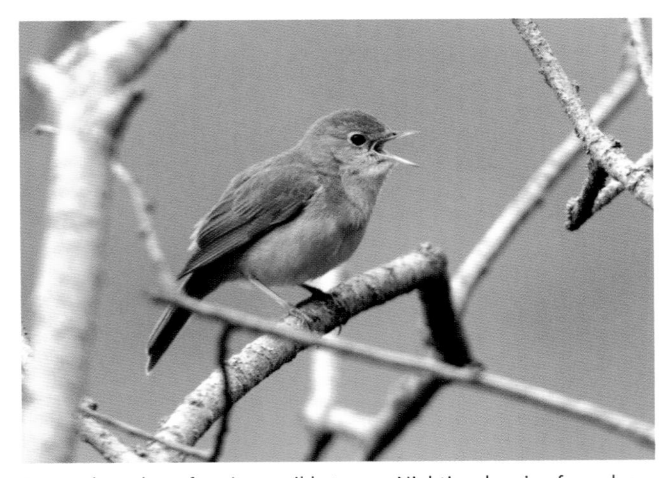

Easy to hear, but often impossible to see, Nightingales sing from deep cover.

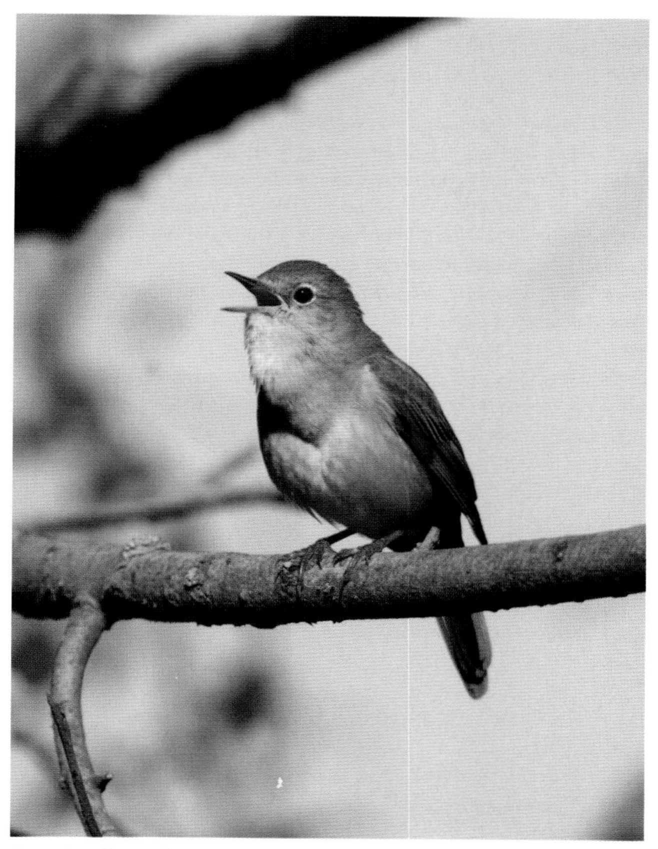

Occasionally Nightingales sing out in the open.

DIET Feeds mainly on insects and worms, and also on fruits. It is only likely to be seen in gardens in suitable habitats, where it may take live mealworms, especially when it has young to feed. It may also visit a pool or birdbath to drink and bathe.

NESTING The nest is usually low near the ground in dense cover. The female lays one clutch annually of 4–5 glossy olive-brown eggs. They are incubated by the female, and both adults feed the young.

BLACK REDSTART

Phoenicurus ochruros

The Black Redstart is perhaps not necessarily known as a garden bird, but it does have a preference for nesting in cracks and crevices in buildings. It began to increase its range into Britain during the years after the Second World War, exploiting bombed-out buildings as nesting sites following the Blitz. The male is a smart chat, similar in size and posture to the Robin. It has a grey crown with a white brow, a black face, breast and head, a slate-grey mantle and a bold orange-brown belly, flanks, rump and tail. The tail is constantly flicked to reveal two brighter red-brown central feathers. The female is greyish-brown with a reddish-brown lower rump and tail. There are several subspecies that vary slightly in colours and markings.

LENGTH 14cm.

HABITAT AND DISTRIBUTION This is a widespread species found across Europe and Asia and into northern parts of Africa. It is not very common in the UK and can be seen in urban sites and on cliffs in most months of the year; numbers increase during passage.

DIET Black Redstarts are insect eaters and take mealworms and wax worms if these are provided for them. The natural diet is flies, which they catch in mid-air, and small larvae and crustaceans. They can often be seen feeding on the tide line around coastal sites.

NESTING Often nests in wall cavities.

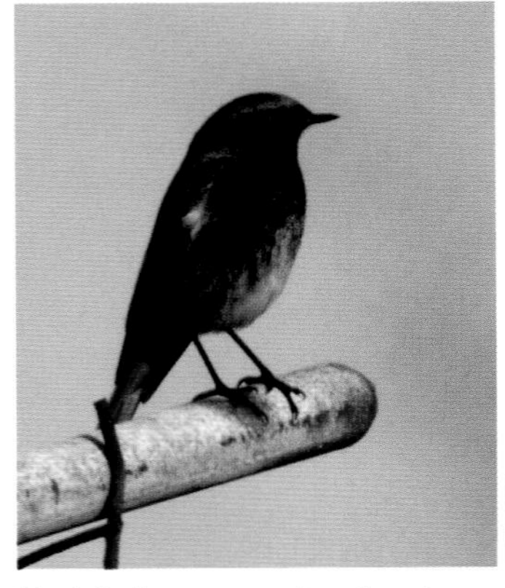

Black Redstarts are often found on industrial sites.

The female is a paler grey all over, but the red tail is distinctive on both sexes.

COMMON REDSTART

Phoenicurus phoenicurus

A close relative of the Robin, and very similar in size and shape, the summer male Common Redstart even has a red breast. It also has red flanks, a red rump and a red tail. Its face is black and it has a grey crown, nape and back, a pale grey-and-white forehead, and black legs and bill. It is a very smart-looking bird that constantly flicks its tail when perched. Females are greyish-brown above and orange-buff below. They also have a red tail, but it is a lot less brightly coloured than the male's. Both adults are less strongly marked in their winter plumage. Males sing a squeaky, warbling song interspersed with a soft *hweet* or *tick* to attract a mate.

LENGTH 14cm.

HABITAT AND DISTRIBUTION A summer migrant to parts of the UK and mainland Europe where the old oak and birch woodlands it prefers are situated, usually arriving in April.

DIET Feeds on insects, larvae and spiders. In autumn it feeds on berries before migration to the south. Although the birds are usually quite shy, they can be attracted to a garden with mealworms.

Females and winter plumage males are less boldly marked.

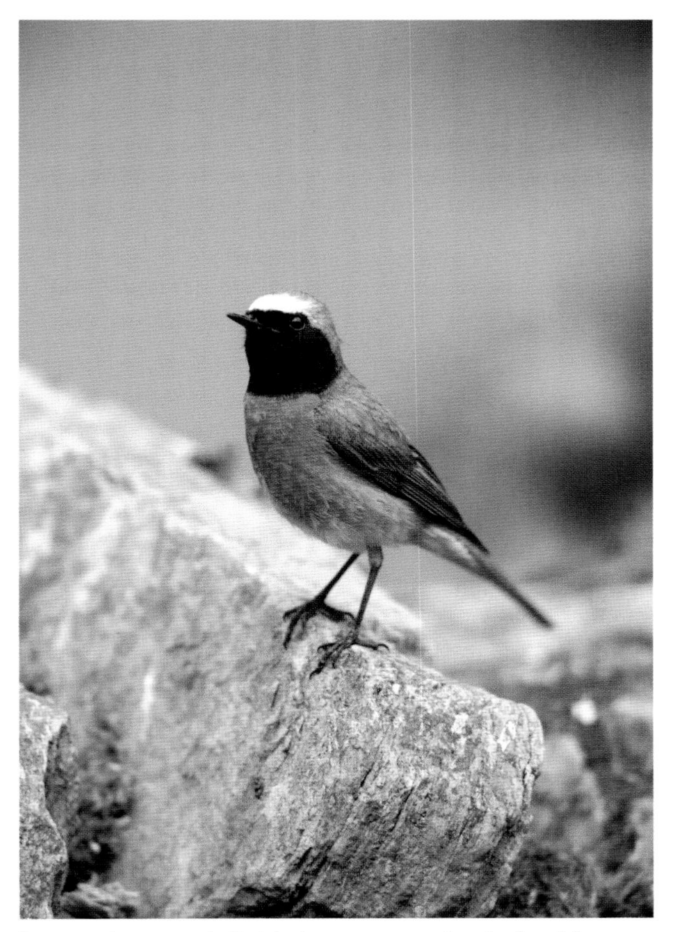

Summer plumage male Redstarts are very smart and colourful.

NESTING Once paired the birds find a nest site in a hole or crevice in a tree or dry-stone wall. They may also be attracted to an open-fronted nest box. The female builds the grass-and-moss, cup-shaped nest, lining it with hair and feathers for her glossy pale blue eggs. Once the eggs hatch the male joins in with feeding the chicks.

STONECHAT

Saxicola rubicola

Named after its alarm call, which sounds like two flint stones being knocked together, this chat is a close relative of the Robin and redstarts, and similar to them in shape and mannerisms. The colours, however, are different. Its back and wings are brown, and there are white patches on the neck and rump, and rusty-orange on the breast and underparts; in summer the male has a black head. Females are similar but duller in colour, have less defined white patches and do not have a black head. Juveniles are like females but slightly duller. Stonechats do not have a supercilium.

LENGTH 12cm.

HABITAT AND DISTRIBUTION Stonechats are found in many habitats, such as dunes, coastal scrub, heath and moorland. The UK population increases in winter, when some continental birds arrive.

DIET Feeds mainly on insects; also worms and spiders. Stonechats are not generally garden birds, but may visit gardens situated in the kind of habitat that they prefer, especially if live mealworms are served regularly.

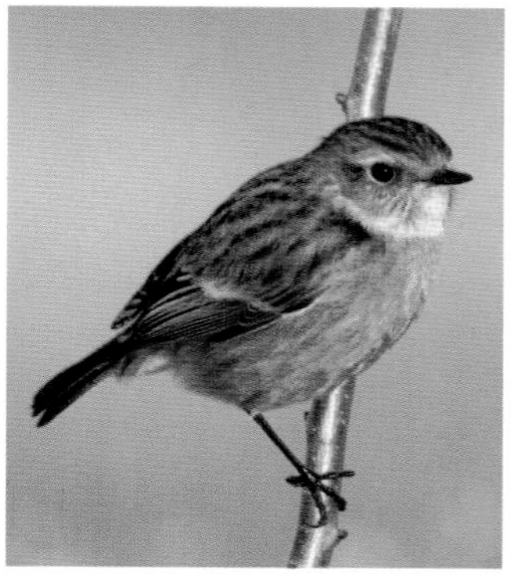

The female Stonechat has a more subdued plumage.

NESTING The nest is typically made from grass, lichen and moss, and is built by the female. The eggs are laid on a cosy bed of hair and feathers. The 3–8 pale greenish-blue eggs with fine red-brown speckling take 15 days to hatch. The male helps with feeding the young on a wide diet of caterpillars, flies, moths and other insects.

This male Stonechat is in a typical pose atop tall vegetation.

WHEATEAR

Oenanthe oenanthe

Another close relative of the Robin, the Wheatear is a summer visitor to the UK and is most likely to be seen in gardens during migration, particularly in areas near the coast or close to its preferred upland habitat. The same size and shape as a Robin, the Wheatear has a smart appearance. It has a blue-grey back, nape and crown, black wings, a white belly, and a peachy-orange flush to the breast, throat and lower part of the cheeks (the Greenland race has the peachy colour on the belly too). The eye is within a broad black band that has white stripes above and below. The best identification feature is the inverted black 'T' on the tail and the brilliant white rump, most visible when the bird flies away from you. The female is browner and less strongly marked, and by autumn after moulting the male looks similar to her.

LENGTH 15cm.

The summer male is boldly marked and has a peachy-coloured breast.

HABITAT AND DISTRIBUTION Mainly upland hills, pastures and cliffs. Wheatears are very much ground-dwelling birds, although they do like to perch on high objects such as rocks and tussocks, and are often seen on dry-stone walls. They are summer visitors to much of Europe, wintering in Africa.

DIET The diet consists mainly of insects. It may come to live mealworms in a garden (and become protective of the food supply).

NESTING Nests on the ground in Rabbit burrows and holes under stones, as well as gaps in stone walls. Typically lays 3–8 eggs but sometimes more.

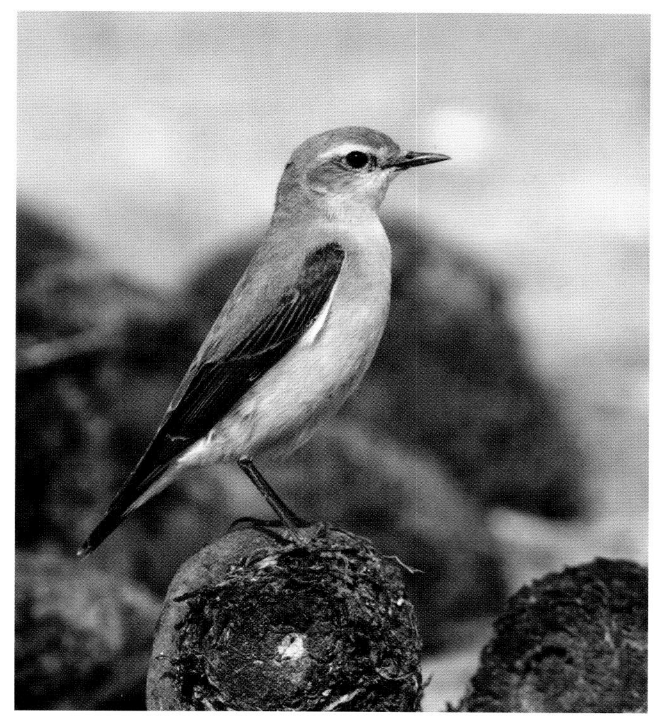

A female Wheatear standing erect as it looks for flying insects.

DUNNOCK

Prunella modularis

Formerly named Hedge Sparrow, the Dunnock should more properly be called Hedge Accentor. Although it is sparrow-like in colour and habits, it is an insectivorous bird. A close look at it reveals its attractions – a steely grey-blue head and breast, two-tone streaky brown back and flanks, and pink legs and feet. The sexes are similar, with the females being slightly duller. The song is a pleasant but tuneless jangle of notes, usually sung from a high perch on a bush, tree or garden fence. It also has a sharp *tseep* contact call. An old name for the Dunnock was Shufflewing because of the repeated wing-flicking movements it makes as it skulks about.

LENGTH 14cm.

HABITAT AND DISTRIBUTION A numerous and widespread species that crops up in many varied habitats, such as gardens, parks, open woodland, heaths, farmland and hedges across most of Europe. It is, however, on the Amber List due to declining numbers since 1980, particularly in woodland, possibly because of changes in habitat management. It is faring better in gardens, so helping it in the garden is a good thing to do.

DIET The sharp, slender bill is most suited to prising insects from crevices, but the species does eat seeds and a variety of berries and vegetable matter. In the main it is a ground feeder and is quite inconspicuous, searching out food under vegetation.

NESTING Breeding is a bit more complex than that of most birds. The female builds a nest of twigs lined with grass, moss and hair. She lays and incubates her glossy blue eggs, and her mate helps with feeding the young. However, Dunnocks are very promiscuous and few nests contain a clutch of eggs that are all fertilised by the same male, and most males mate with several females in the same breeding season. There may be 2–3 clutches of 4–6 eggs each year, a strategy that guarantees a wide bloodstock of birds.

Male and female Dunnocks are similarly marked.

HOUSE SPARROW

Passer domesticus

The House Sparrow is mainly several shades of brown, the male having a richer chestnut shade on its back and head, and a grey crown. The bill is conically shaped, typical for seed eaters. The undersides are mainly off-white and grey. Females are a paler and duller brown shade overall. There are several species which the House Sparrow can be confused with, and there is some variation in the colour and markings of the 12 subspecies.

House Sparrows are very social and small flocks gather to feed, dust bathe, drink and wash together. They are usually quite noisy, voicing happy-sounding chirrups. Ringing records show that they are not very mobile, usually venturing not more than a few kilometres other than in hard weather conditions. Being gregarious birds, they are prone to several diseases, which are usually spread as a result of poor hygiene at feeding or watering sites. They have declined as a result of this and other factors in recent years.

LENGTH 15cm.

This adult male is perched on a garden hedge.

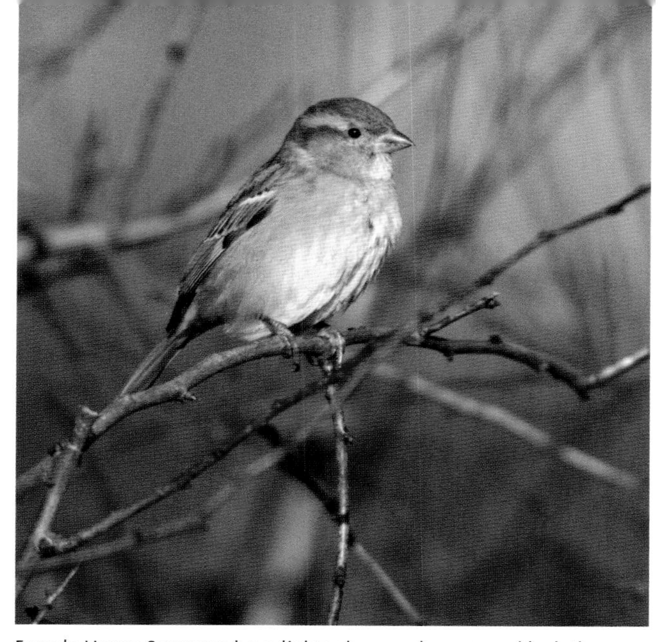

Female House Sparrows have lighter brown plumage and lack the black bib and grey crown.

HABITAT AND DISTRIBUTION Closely associated with humans and found around houses and cultivated areas, House Sparrows occur right across the world. They originally ranged from Europe and Asia into China, but European settlers introduced them to Australia, America and South Africa as a reminder of home.

DIET The diet comprises a high percentage of seeds, but they also eat insects, flies and caterpillars. They are adaptable feeders and take leftover human food in urban areas. They may be locally very common in gardens, yet a few kilometres away be very scarce.

NESTING Nests in cavities in trees and buildings. The male occupies a site and starts calling females from a nearby perch, often building the nest before a mate is found. The female lays 4–5 white eggs with blue-grey spots. The young leave the nest about five weeks later.

TREE SPARROW

Passer montanus

Tree Sparrows are just slightly smaller than House Sparrows, which they may be confused with. The male and female are very similarly marked with a 'milk-chocolate' brown crown, a white collar and a black cheek-patch. The back has several rich tones of brown with buff fringing on some of the darker feathers; the bill is black with a yellow base. In flight two wing-bars are visible – House Sparrows have only one. Tree Sparrows are generally sedentary and small flocks stay within a territory in suitable habitats.

LENGTH 14cm.

HABITAT AND DISTRIBUTION Farmland and suburbs; rare in cities. In winter flocks feed in stubble with finches and buntings. The species has a massive breeding range across most of temperate Europe and Asia. It is common in some parts of its range, but decreasing in the UK, where there has been a population loss of more than 50 per cent since the 1970s. Some northern birds move south in winter, and the UK sees a slight increase in numbers at this time.

Tree Sparrows are richer-coloured than House Sparrows and have a chestnut crown.

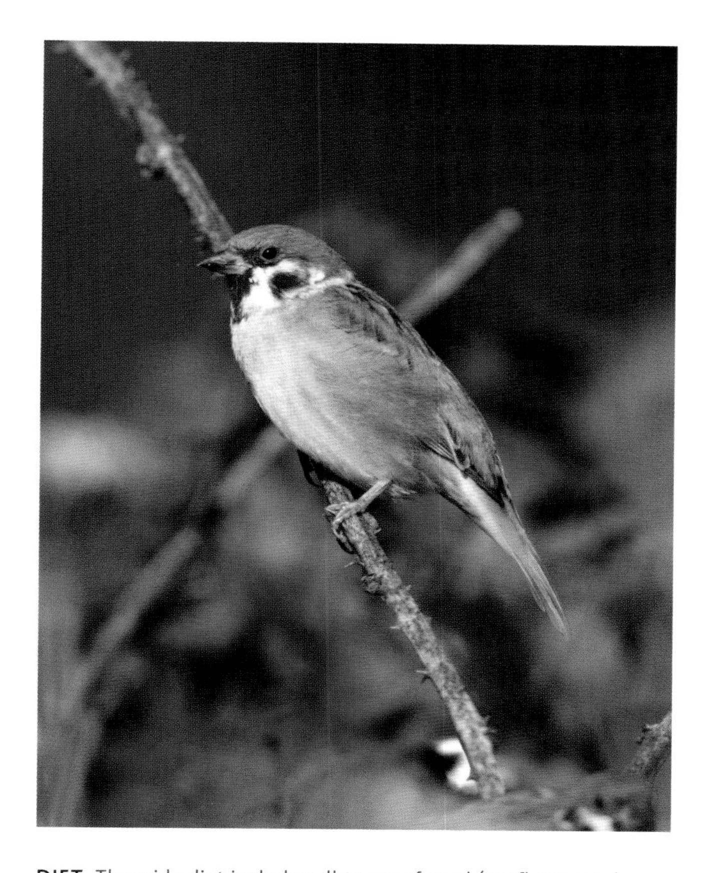

DIET The wide diet includes all types of seed (sunflower and millet in particular). The birds also eat insects and take live or dried mealworms in rural gardens.

NESTING Mates for life and usually breeds in small colonies. The nest is made in a cavity in a tree or building. The birds also use nest boxes very successfully, and projects with boxes in some areas seem to have slowed or halted the population decline. They lay 3–9 eggs, which are glossy white with brown marks. Both adults share incubation and feeding duties. There may be three broods a year, but two are more normal.

YELLOW WAGTAIL
Motacilla flava

An attractive summer visitor to Britain, the Yellow Wagtail glows yellow like the summer sun and produces a happy-sounding *tsweeep* call. It is a confusing species in that there are (depending on the source) 15–20 recognised subspecies and a few that are hybrids between them. The male is bright sulphur-yellow underneath and olive-yellow to olive-brown above, with some black and brown showing on the open wing. The subspecies found mostly in the UK has yellow cheeks and an olive-yellow crown, which on some birds can be quite bright yellow. Other races have blue-grey, black or grey heads, and a mix of pale white, yellow, black or grey supercilium and/or moustachial stripes. Females are slightly duller with a browner back and a grey wash to the face and head. The Grey Wagtail is also partially yellow, which can add to the confusion in identifying the species – however, once you have seen a Yellow Wagtail identification becomes easier to get to grips with.

LENGTH 16cm.

Female Yellow Wagtail collecting nest material.

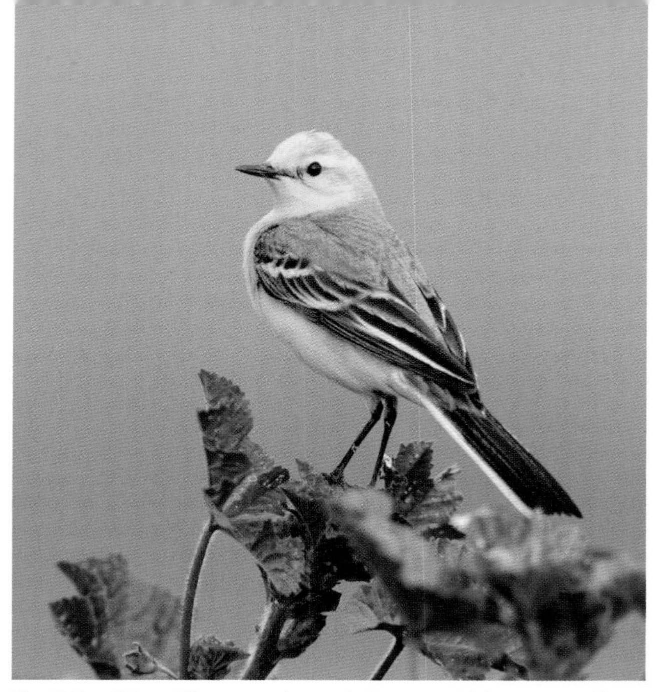

The Yellow Wagtail has several races in Europe, which are generally similar apart from head colour and markings.

HABITAT AND DISTRIBUTION Damp meadows, farmland and paddocks in much of Europe. The birds also frequent dung heaps, especially on migration, when it is easy to view several at one site. They migrate to Africa for the winter.

DIET Feeds on all sorts of insect life, often right under the feet of horses and cows, snapping up the bugs disturbed by the mammals' hooves. In gardens it feeds on mealworms and can become very tame, feeding from the hand. It may visit gardens in areas adjoining the habitats it prefers, and can show up perched on hedges and at water.

NESTING Nests on the ground on a tussock or in arable crops. The nest is a grassy cup lined with animal hair, in which 4–7 glossy buff eggs with grey-brown speckles are laid. The female incubates them and the male joins in with feeding the young.

GREY WAGTAIL

Motacilla cinerea

The blue-grey upperparts, black-and-white face, and yellow breast, belly and rump, combined with a constantly wagging tail (the longest tail of all the wagtails), make this a very attractive bird. Usually found near running water, the Grey Wagtail is a very agile flyer and runner, darting about and catching the flies that make up the majority of its diet. In winter the birds are paler in colouration, but essentially retain the same marks. The only confusion species is the Yellow Wagtail, but this is a summer visitor only, whereas the Grey Wagtail is resident throughout the year.

LENGTH 19cm.

HABITAT AND DISTRIBUTION Vicinity of water, including streams, rivers, lakes and ponds, throughout much of Europe. Although the species is not a true migrant, some birds move south in bad weather, and they also move to lower altitudes in the winter months.

Grey Wagtails prefer to be near running water.

DIET Insectivorous, often chasing insects across water. It is most likely to be seen in a garden if there is a medium-sized pond or river nearby. It struts about on water-lily leaves and flits around muddy edges while hunting insects, and also looks for food on lawns.

NESTING Nesting takes place in April–June. The nest is usually in a crevice in a wall, and is lined with moss and hair. There may be two broods a year, each of 3–7 creamy-buff eggs with darker spots. The female incubates the eggs and both parent birds feed the young.

As the name suggests Grey Wagtails constantly wag their long tails.

PIED WAGTAIL

Motacilla alba

This is a long-tailed, black-and-white bird that is constantly on the move, wagging its tail up and down and running around on 'clockwork-toy' legs as it catches its insect prey. In the UK the Pied Wagtail in the most common subspecies, but the nominate White Wagtail is frequently seen in eastern areas of Britain. It has a grey back and rump with a sharp line on the nape between the grey and the black on the head, whereas the Pied Wagtail's back is black.

LENGTH 18cm.

HABITAT AND DISTRIBUTION Pied Wagtails are commonly found in damp open spaces. They favour large, short-grass places, and in towns generally occur on paved areas, often near water features. There are 11 subspecies within this wagtail's range, which is quite large, stretching from Iceland across to Japan and China, down into India and South-east Asia and across North Africa.

Juvenile Pied Wagtails are pale with some buff yellow on the head and breast.

British Pied Wagtails have black backs.

DIET Insects are the main food. In gardens the birds take mealworms, wax worms and anglers' maggots, often becoming possessive of the food and chasing other birds away.

NESTING Nests in holes and crevices, frequently in buildings, and in open-fronted nest boxes. The nest is made of grass and moss, and the female lays 3–7 eggs, often twice in a season. Both adults incubate the eggs and feed the young.

MEADOW PIPIT

Anthus pratensis

This mundane little brown bird is difficult to see, being streaky grey-brown to olive-brown above and streaky-buff below. However, on close viewing it appears quite appealing. It is a complicated bird to identify because there are several confusion species, all of about the same size, and there is some colour variation within the species itself. The best places to see Meadow Pipits are high vantage points where they perch, preen, watch for predators and sing from; they often perch up and face into the strong wind in moorland areas.

LENGTH 15cm.

HABITAT AND DISTRIBUTION Pasture, moorland and arable farmland are the preferred habitats; in winter they move to other habitats, like saltmarshes. They occcur across Europe.

Despite the dull colours Meadow Pipits are attractive birds that resemble a small thrush.

Meadow Pipits are variable in colour; this one is quite brownish.

DIET Mainly eats insects including small flies, adding the seeds of several grasses and heather to its diet in winter. It may come to gardens in suitable habitats, particularly in winter when food is scarce.

NESTING The nest is constructed in dense vegetation on the ground. When arriving at their nest the birds usually drop down away from it and walk to it under cover. Despite their efforts to conceal the nest, they still get parasitised by Cuckoos – a fall in Meadow Pipit populations will thus have a knock-on effect on Cuckoos.

CHAFFINCH

Fringilla coelebs

This is a familiar bird to most people because it is quite tame and turns up at places like picnic sites and tea gardens to peck for scraps and crumbs of food. Because of its familiarity it is often overlooked, but it is a pretty passerine with a pastel peach-orange breast, brown back, black wings, white wing-bars and greenish-grey rump; in flight it has noticeable white outer-tail feathers.

LENGTH 15cm.

HABITAT AND DISTRIBUTION All types of woodland, parks and gardens. This is the second most common bird in the UK and is widespread throughout Europe.

DIET Along with Blue and Great Tits, this is one of the first birds to exploit new feeder sites. It feeds mainly on seeds, but supplies its young with insects too. Chaffinches take a wide variety of bird-table food such as sunflower, millet and niger seeds, and peanuts; they also take live and dried mealworms and wax worms.

The female has similar markings to the male, but lacks the bold colours.

NESTING Nests in the forks of trees or in hedges and shrubs. The female builds a neat little moss-and-grass cup with bits of bark and lichen held in place with spiders' webs; this is lined with hair, wool and feathers. She lays 4–5 eggs, or sometimes more, which are glossy pale blue with purple-brown blotches. Incubation takes 12–16 days and the fledglings leave the nest after 14–18 days.

A very smart summer male Chaffinch.

BRAMBLING

Fringilla montifringilla

Male Bramblings are very distinctly marked with a dark head, orange breast, white belly and dark speckles on the flanks. The clean white rump distinguishes them from the similar Chaffinch, which has a dull grey-green rump. Females and juveniles are more difficult to tell apart from female Chaffinches, but the white rump is always diagnostic, as is the bill colour – that of the Brambling is yellow, and black on breeding males, while the Chaffinch's bill is grey or dull greyish-pink.

LENGTH 15cm.

HABITAT AND DISTRIBUTION Bramblings are entirely migratory so winter is the time to see them – they occur in large flocks especially where there are Beech trees. They are widespread across northern Europe, and mainly winter visitors further south and in the UK.

Adult male Brambling moulting from winter to summer plumage; the head is completely black in summer.

Female Bramblings can be confused with Chaffinches.

DIET Beechmast is the main winter diet. Bramblings visit gardens and come to feeders. Mixed seed for finches is a good food for them, and they also like sunflower hearts. Scattering food on the ground attracts them as they are most naturally ground feeders. At feeder sites the numbers vary – in some years there may be none or perhaps only one or two birds, while in others there may be as many as 40.

NESTING The nest is a deep cup of moss, grass and hair, lined with feathers and wool, and decorated with lichen and bark. It is usually built in the fork of a tree.

GREENFINCH

Carduelis chloris

This medium-large finch is, as its name suggests, mainly green. It is stocky in build and has a distinctly forked tail. Males are a richer green than females, which are more grey-green with streaky upperparts. Both sexes have yellow – brighter in males than in females – on the outer edges of the primaries and tail feathers. The legs and bill are a flesh-pink colour. Males utter a repeated *dizwheee* song in a wheezy voice from a high perch in spring. Greenfinches have a bounding flight, and call in time with the undulations.

LENGTH 15cm.

HABITAT AND DISTRIBUTION Woodland edges, open woodland, parks, gardens and farmland with hedges, year round across Europe. After breeding the birds disperse and move around in flocks, with northern birds moving down to swell numbers in the south during the winter months. Greenfinches have declined considerably in the last five years because they have been dying from trichomonosis, a parasitic disease that lives in the digestive tract and closes up the throat, affecting finches and doves especially.

A summer male at a sunflower kernel feeder.

DIET Eats seeds and berries, as well as some insects during the breeding season. Frequently comes to feeders and stays to feed for long periods, favouring peanuts and sunflower seeds; if a feeder contains mixed seed it will turf out what it does not want and drop it to the ground in order to get at the seeds it prefers. If there are signs of birds with trichomonosis (see p. 168) in a garden, it is a good idea to stop the feeding and empty birdbaths for a few weeks – the parasite cannot survive long outside a host. Before resuming feeding all containers should be thoroughly cleaned, disinfected and rinsed.

NESTING Usually nests in loose colonies in dense shrubs. The female builds a cup structure from twigs and dried grass lined with moss and hair. She lays 3–7 buff-coloured eggs and incubates them.

Females and juveniles are a less vibrant green colour.

GOLDFINCH

Carduelis carduelis

This delightfully happy-sounding, brightly coloured and highly sociable finch is always a joy to see. It is unmistakable, with a red face, white cheeks, black head and vivid yellow wing-patches. Even the duller females and juveniles have yellow on their wings, so they are quite easy to identify. The birds' liquid, twittering songs and calls are also very distinct. They almost always occur in small flocks of loose family parties.

LENGTH 13cm.

HABITAT AND DISTRIBUTION Farmland, heaths, orchards and gardens in most of Europe. Goldfinches undertake some migratory movement, and there are increased numbers in the southern parts of their range during winter.

Juvenile Goldfinches still have that diagnostic yellow wing bar.

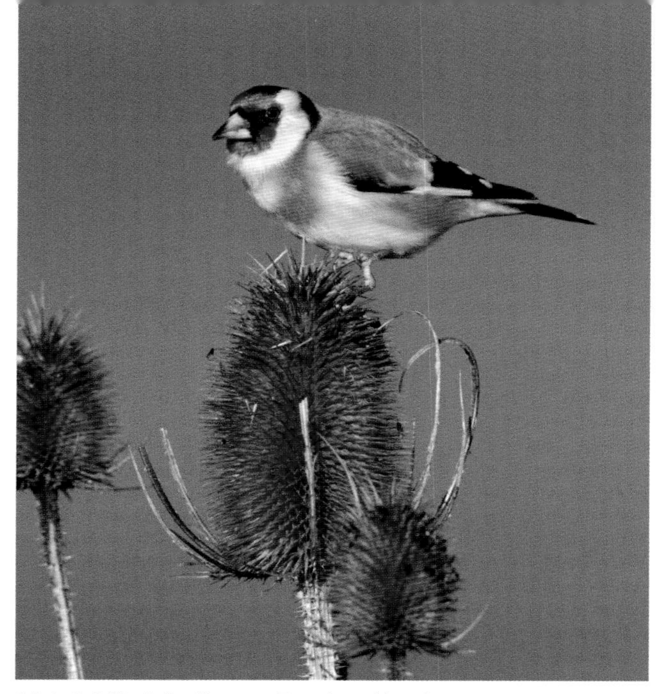

Adult Goldfinch feeding on a Teasel seed head.

DIET The bird's long, sharp bill is perfect for feeding on the seeds of spiky plants; it also eats insects during summer. Often seen feeding on Thistle and Teasel seeds, its extensive diet also includes Dandelion, alders and birches. Goldfinches are increasingly frequent garden-feeder visitors. They favour oil-rich, high-energy seeds such as niger and sunflower kernels, and may also take mealworms, live or dried (dried ones should be soaked in warm water before they are put out). Garden food may be necessary to Goldfinches due to the decline of their natural food in the countryside; the loss of hedgerows as well as verge cutting before the plants go to seed could be a factor in their decline.

NESTING Breeding begins in April. The nest is constructed from moss and grass lined with wool, and there are as many as three broods annually, each usually comprising 3–5 glossy pale blue eggs with reddish-brown spots. The young hatch after about 12 days and leave the nest 18 days later.

SISKIN

Carduelis spinus

This very small finch is about the size of a Blue Tit, and like this species it is very agile, being able to hang upside-down on flimsy branches to feed on small seeds. The striking-looking male is yellow with a hint of green. He has very bright yellow cheeks, rump and wing-bars; the belly is yellow streaked with dark lines, and the crown and bib are black. Females are duller without the black on the head, and can be confused with the larger Greenfinch at a distance; juveniles are a plainer brown with heavy streaking. All Siskins have a deeply forked end to the tail.

LENGTH 12cm.

A female taking a drink.

HABITAT AND DISTRIBUTION Coniferous and mixed forests in winter in much of Europe.

DIET Siskins have small, pointed bills that are perfectly designed to extract seeds from the small cones of alders, birches, spruces and pines, which form the main diet. When food is short they come to garden feeders and feast on niger seeds, fat and peanuts – the birds are especially attracted to red-coloured feeders. They also come to drink and bathe, often in pairs or family parties. Once rare in gardens, they have become regular winter visitors in many areas.

NESTING The nest is constructed high in a conifer tree by the female. It is made up mainly of twigs and grass lined with hair. She lays 2–6 glossy pale blue eggs with pinky-purple spots, and like most finches incubates them alone. The male helps with feeding the young.

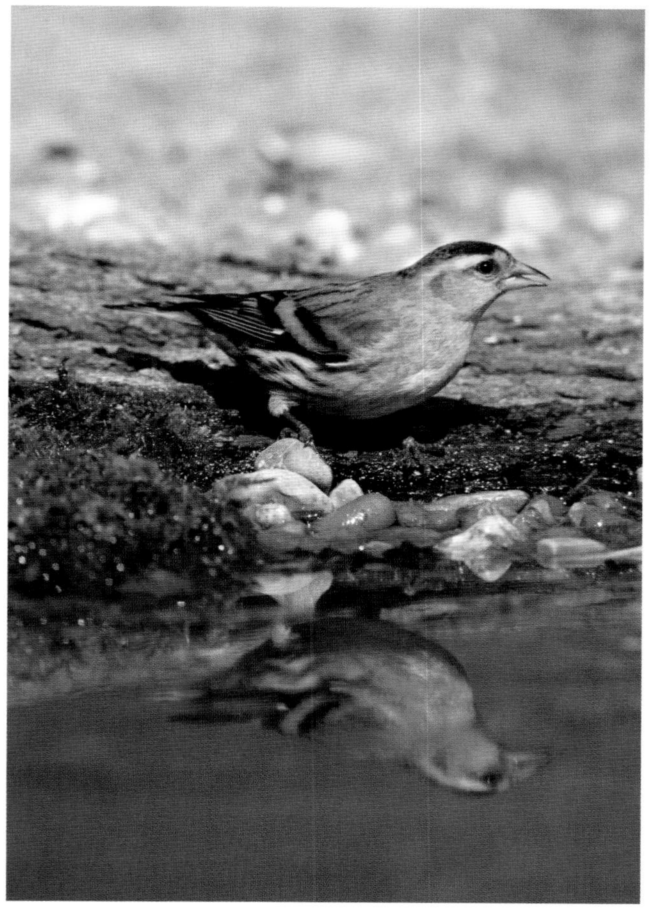

A male Siskin in summer. Siskins can be attracted to gardens with water and niger seed.

LINNET

Carduelis cannabina

This is one of the smaller members of the finch family. Redpoll and Twite are slightly smaller and can be confused with it, but both sexes of the Linnet have a chestnut-brown mantle and whitish underparts with brown streaks, the female being duller than the male. Both have a grey bill, flesh-brown legs and a deeply forked tail. Summer-plumage males are more distinctly marked in that they have a pale crimson breast and forehead, a grey head and streaked markings on the throat; females are duller. At this time of year the male likes to find a high, open perch to sing from, the song being a pleasant medley of wheezy chattering notes. Linnets make a twittering call in flight, similar to that of Greenfinches.

LENGTH 13cm.

HABITAT AND DISTRIBUTION Open fields with bushes and wasteland, and farmland and coasts in winter. Widespread across much of Europe, but declining due to changes in agricultural practices.

DIET Linnets eat all kinds of small seeds, as well as some insects. In winter they form up into flocks, often with other finch species. They come to water to drink in gardens, as well as to seed feeders, usually favouring niger seed and sunflower hearts.

Linnets are unusual in most gardens, but sometimes visit a pool or birdbath.

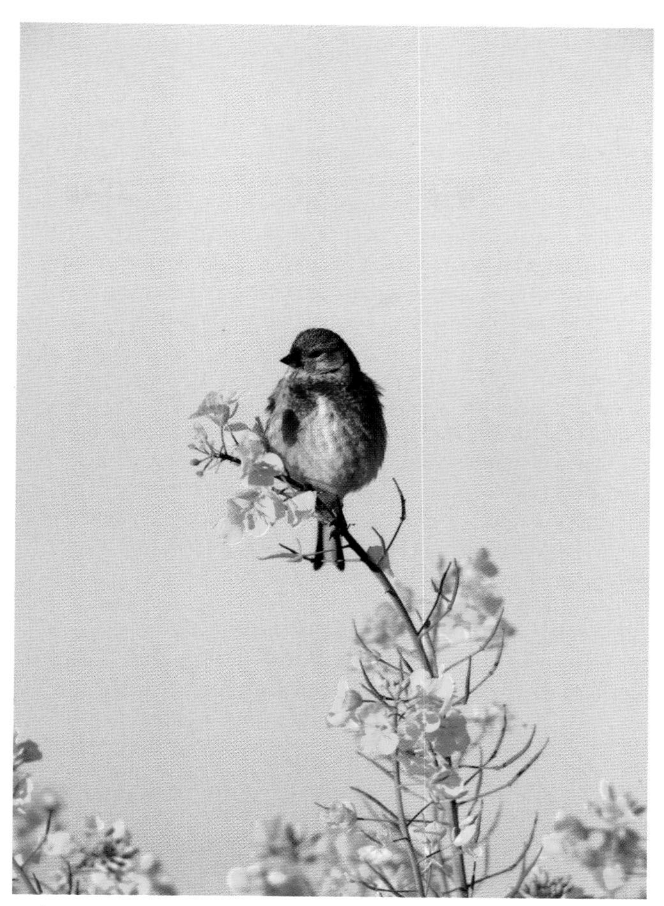

Male Linnets usually sing from a high perch.

NESTING Nests in densely twigged or thorny vegetation. On heathland the birds have a preference for Gorse. In gardens and elsewhere they use hedges and tangled bushes. The nest is a grassy cup lined with moss, hair and wool. The female lays 4–6 pale blue eggs with purple-brown speckles. The female incubates the clutch and both birds feed the nestlings. In some years Linnets may have as many as three broods.

LESSER REDPOLL

Carduelis cabaret

One of the smaller members of the finch family, the Lesser Redpoll is greyish-brown with darker brown streaking above, and a lighter buff-white below with whiter flanks and brownish streaks; there are two buff-coloured wing-bars. The tail is brown and has a fork-shaped end. The red forehead and little black bib help to tell it apart from the similar Twite *Carduelis flavirostris* and Linnet. The summer plumage includes a pink flush on the breast and rump. Females are similar but without the pink.

LENGTH 12cm.

HABITAT AND DISTRIBUTION Breeds in forests (especially birch) and on heaths from Britain to southern Scandinavia and central western mainland Europe. This species is mainly sedentary, but birds from the north of the range move south in colder weather. The Mealy Redpoll *Carduelis flammea* is a rare winter visitor to northern and eastern Britain.

DIET Redpolls are common visitors to garden birdfeeders, usually favouring niger and millet. Naturally they feed on alder, spruce and beech seeds, either while acrobatically hanging in the tree branches or down on the ground among the leaf litter. In summer they also eat insects.

NESTING The female builds an untidy cup-shaped nest lined with hair in a woodland scrub location. She incubates up to five pale blue eggs with pale pink markings for about 12–14 days, and both parents feed the young.

Female redpolls are brown and streaky, but still have red on the forehead.

Spring male Redpolls show an extensive pink wash on the breast, but females are much drabber.

COMMON CROSSBILL

Loxia curvirostra

These birds are usually seen as noisy flocks at the tops of coniferous trees. They dart from cone to cone, feeding on the seeds held within, hence the highly specialised, sharp, crossed-over points on the bill mandibles, for which the species is named. This well-built member of the finch family has a large head and a forked end to the tail. Males are a bold orange-red with black on the wings and tail; females are quite bright, having yellow-green plumage and dark green-grey wings and tail. Within the age groups of a flock there is some variation in colour.

Although Common Crossbills are mainly sedentary, they are irruptive in behaviour and may appear in good numbers at a site in one year but not in the next, moving around to where food is most abundant.

LENGTH 16cm.

HABITAT AND DISTRIBUTION Fir and spruce forests in much of Europe.

The female Crossbill is greenish with a yellowish-green rump.

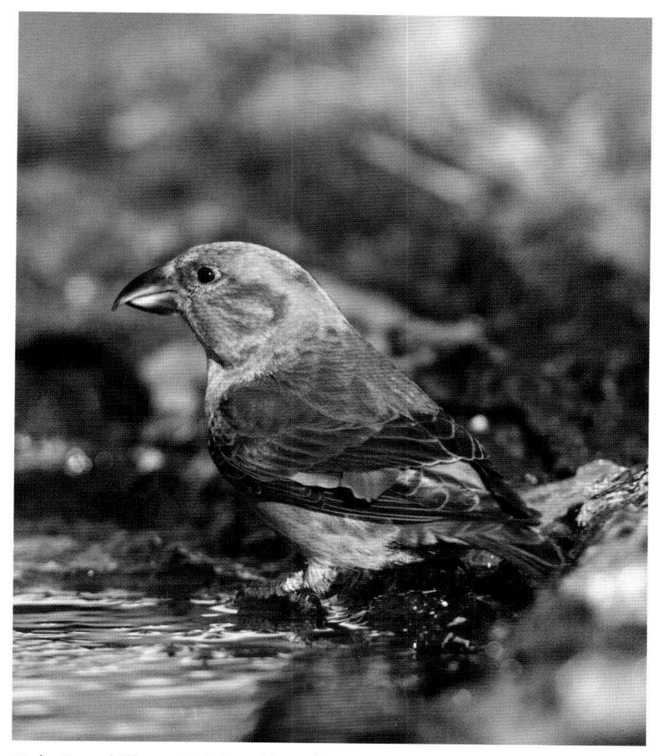

Male Crossbills are brick-red in coloration, sometimes with yellow-orange on the rump.

DIET Despite the specialised bill for extracting coniferous seeds, mainly from spruces, larches and pines, Common Crossbills have a varied diet that also includes other seeds, insects and berries. Because of its diet this is not a bird that comes to birdfeeders, but it does come to water very frequently to drink and bathe. It tends to bathe once a day in around the middle to the late afternoon.

NESTING Nests high in evergreen trees, where the female lays four pale blue eggs with purple spots in a cup made from twigs and lined with moss, grass and hair. The young leave the nest about five weeks later.

BULLFINCH

Pyrrhula pyrrhula

This is a stunning-looking and unmistakable species. The male's bright pink-red breast and cheeks, grey back, white rump, and black head and tail stand out well. Females and youngsters are marked like the male, but with the colours desaturated so that the breast and cheeks are a buff-pink.

LENGTH 16cm.

HABITAT AND DISTRIBUTION Once considered a pest because they feed on the buds of fruit trees, Bullfinches were persecuted causing a reduction in population. They can be found right across Europe, but northern birds migrate south in winter. There are nine races that although similar do slightly differ in size and colour.

DIET Bullfinches can be irregular visitors to feeding stations, but when natural food is scarce they come to feed on seeds and suet cake. Provision of water is a better means of attracting them as they drink several times a day.

NESTING Bullfinches breed in mature scrub and mixed woodland that is generally above 4m high. The female builds a loose, cupped nest of twigs lined with moss and hair. She lays 4–7 eggs, and both adults feed the young.

Female Bullfinches have beige-pink plumage.

The male Bullfinch is unmistakable, but only shows his white rump in flight.

HAWFINCH

Coccothraustes coccothraustes

This is a big, bull-necked finch with a huge, powerful bill and a short tail. It is brown on the back and has an orange-brown belly and breast with a richer orange-brown head, and a black eye-stripe. It has white wing-bars and a white tip to the tail – striking markings that make it easy to identify in flight as it darts directly past at speed.

The Hawfinch is shy and wary, and is not an easy bird to see. When feeding on the ground Hawfinches hop, and they are dominating and aggressive to other birds, their own kind or other species. They are, however, quick to take flight at the slightest sound or movement.

LENGTH 16.5–18cm.

HABITAT AND DISTRIBUTION Found right across Europe and into Asia, the birds' preferred habitat is deciduous and mixed woodlands with Hornbeams, the trees they usually choose to nest in. Hawfinches are mainly sedentary in the UK, but continental birds are more mobile, with the European and Asian populations migrating north to south. Some birds come to Britain, almost doubling the UK numbers in winter. The species is unfortunately declining quite alarmingly in numbers.

DIET Feeds mainly on seeds and fruit kernels, including cherries; the birds use their big, powerful bills to crack these open. At times of the year when there are no seeds they eat insects and buds. They generally feed high up in trees. Hawfinches do visit gardens that are within the type of habitat they use, and take peanuts in some gardens in Europe. Cherries, Hornbeam and beechmast may attract them to gardens.

NESTING Breeding starts in April and – as is the case with many finches – the female incubates the eggs and both birds feed the young.

In Britain, a Hawfinch would be an incredibly unusual garden visitor. This beautiful portrait was taken in the Forest of Dean.

YELLOWHAMMER

Emberiza citrinella

The male of this bunting is yellow below and brown above, with a beautiful chestnut-coloured rump. The bill is silvery-grey and the legs are flesh-pink. The tail has a 'V'-shaped notch at the end and the outer feathers are white. Females are similar but duller overall. The song is a trilling sound delivered from an open perch or gate post, *A little bit of bread and no cheese*.

LENGTH 16cm.

HABITAT AND DISTRIBUTION Yellowhammers are found in most open countryside and on arable farmland in much of continental Europe and the UK. The population has dropped by about 50 per cent in the last 25 years due to changes in farming practices, although the set-aside scheme and game feeding seem to some extent to have slowed or halted the decline.

DIET Feeds on seeds and grain, and some insects in the breeding season. Visits large rural gardens in suitable habitats. Feeds on seeds, and may take mealworms occasionally, as well as bathe and drink.

The female is duller and more streaked underneath.

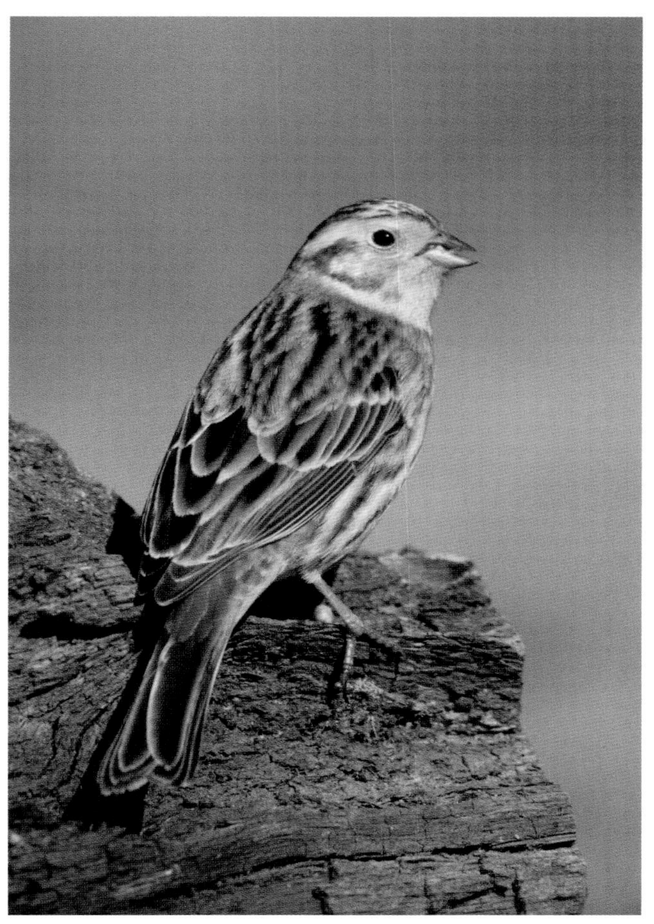

The male Yellowhammer is boldly marked and has a bright yellow head.

NESTING The nest is a grassy cup with moss and hair lining, constructed on the ground. The female builds the nest and incubates the glossy white, faintly purple-grey marked eggs. The brood is 3–6 young with two or perhaps three clutches being produced each year.

REED BUNTING

Emberiza schoeniclus

This bunting is about the size of a House Sparrow and can easily be mistaken for one in winter, particularly in the case of the female. A closer look reveals the difference. The female Reed Bunting is a richer brown with dark streaks; the breast and flanks are streaked, and there is a black moustachial stripe and buff stripes above and below the eye. In flight she shows white outer-tail feathers. The male is generally similar but has a velvety-black head, throat and bib surrounded by a white collar in summer, and a duller brownish version of the markings in winter. Males sings from prominent perches at the tops of bushes or reeds. The song sounds a bit like *tree tree top the tree*, repeated several times in a row.

LENGTH 13.5cm.

HABITAT AND DISTRIBUTION Although this is a species of wet margins and reedbeds, it is also found in open countryside with hedges and grass verges, and additionally seems to frequent dung heaps. It occurs in much of Europe. Reed Buntings are recovering from a decline in numbers that is not fully understood, but may be due to the use of pesticides.

The male Reed Bunting has a black head and white collar.

DIET Feeds mainly on seeds, and also eats insects in the breeding season. Reed Buntings are increasingly coming to gardens in rural areas near their preferred habitats. They will feed on the ground and on a table, and eat most bird foods, including live or dried mealworms (dampened with warm water in the latter case).

NESTING The nest is generally built on the ground in a damp area such as a water meadow or the side of a wet ditch under a hedge. It is a grassy cup lined with hair, wool and feathers. The female incubates the 4–7 dark-blotched, pale olive to lilac eggs, and the male helps to feed the young. A great way to help the birds during the breeding season is to put up a length of rusty barbed wire and pull some sheep's wool onto the spikes – they will come to collect the wool for their nests.

Duller brown colours make the female less obvious, but note the stripy head pattern..

BIRD TOPOGRAPHY

Some of the key terms used to describe the different parts of birds.

FEMALE REED BUNTING

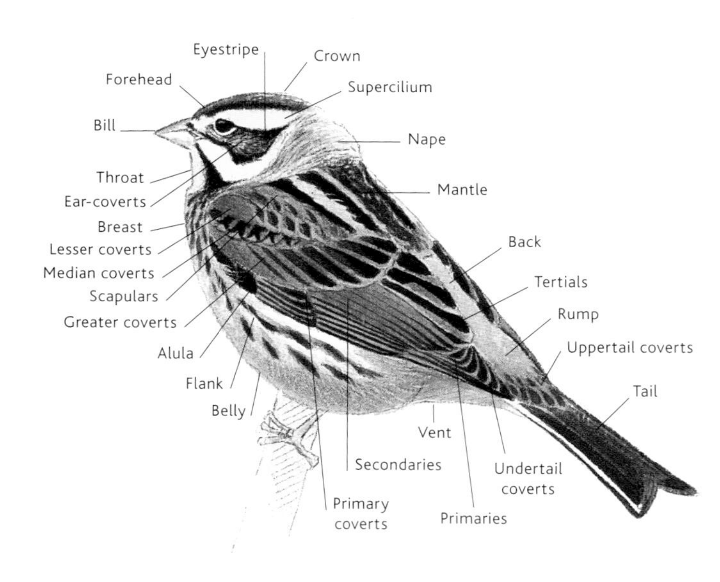

REDWING

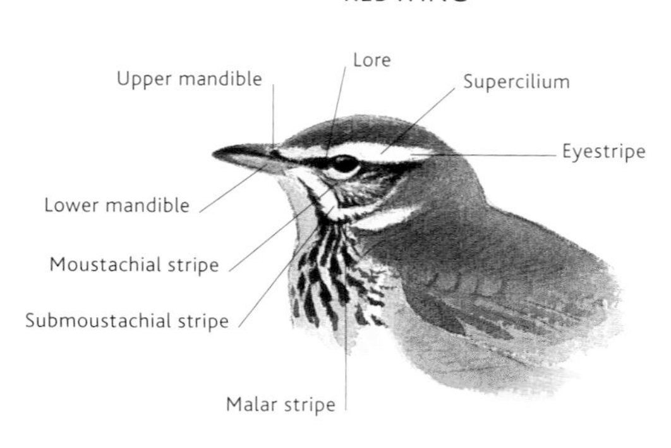

FURTHER INFORMATION

Advice on gardening for birds

General advice: *www.rspb.org.uk/advice/gardening*

Cat prevention: *www.rspb.org.uk/advice/gardening/unwantedvisiiors/cats/index.asp*

Build your own nestbox: *www.rspb.org.uk/advice/helpingbirds/nestboxes/index.asp*

RSPB Wild Bird Foods: *shopping.rspb.org.uk*

CJ Wildbird Foods: *www.birdfood.co.uk*

Conservation bodies

Royal Society for the Protection of Birds (PSPB)
www.rspb.org.uk

Wildlife Trusts
www.wilflifetrusts.org

British Trust for Ornithology (BTO)
www.bto.org

Wildfowl and Wetlands Trust (WWT)
www.wwt.org.uk

INDEX